Neoliberal Ebola

Robert G. Wallace • Rodrick Wallace
Editors

Neoliberal Ebola

Modeling Disease Emergence from Finance
to Forest and Farm

 Springer

Editors
Robert G. Wallace
Institute for Global Studies
University of Minnesota
MN, USA

Rodrick Wallace
New York State Psychiatric Institute
  at Columbia University
Division of Epidemiology
NY, USA

ISBN 978-3-319-82223-5   ISBN 978-3-319-40940-5   (eBook)
DOI 10.1007/978-3-319-40940-5

This Springer imprint is published by Springer Nature
The registered company is Springer International Publishing AG Switzerland

Preface

Disease outbreaks are as much markers of as they are threats to modern civilization. What successfully evolves and spreads depends on the matrix of barriers and opportunities a given society presents its circulating pathogens (FAO 2013; Engering et al. 2013; R.G. Wallace et al. 2015).

For most of its history, *Vibrio cholerae* predated upon plankton in the Ganges delta (Johnson 2006). Only once humanity switched to urban sedentism and later by nineteenth century trade and transport became increasingly integrated in geography and economy did the cholera bacterium evolve an explosive human-specific ecotype.

Simian immunodeficiency viruses, spilling over for centuries, emerged out of their nonhuman *Catarrhini* reservoirs as HIV only when colonial expropriation turned subsistence bushmeat and the urban sex trade into commodities of industrial scale (Wallace 2010; Pepin 2011; Timberg and Halpern 2013).

Domesticated stock served as sources for human diphtheria, influenza, measles, mumps, plague, pertussis, rotavirus A, tuberculosis, sleeping sickness, and visceral leishmaniasis (McNeill 1977/2010; Wolfe et al. 2007). Ecological changes brought upon landscapes by human intervention selected for spillovers of malaria from birds and dengue fever, malaria, and yellow fever from wild primates.

The new pathogens selected for improvements in medical technologies and public health (Watts 1997, Colgrove 2002). In turn, a daisy chain of innovation in agricultural and industrial methods, accelerating demographic shifts and new settlement, rejuxtaposed potential host populations, promoting new rounds of spillover (Kock et al. 2012; R.G. Wallace et al. 2015).

Presently, humanity's organizing ethos orbits neoliberal capitalism, even in opposition (Plehwe et al. 2006). Neoliberalism is a program of political economy aimed at globalizing laissez-faire economics for multinationals, promoting free trade, and shifting state expenditures in favor of protecting private property and deregulating economic markets (Harvey 2005, Centeno and Cohen 2012, Ganti 2014). Once applied at a particular locale, the doctrine has considerable impact on local landscapes and functional ecosystems alike, with decisive effect upon the fortunes of infectious disease (Maye et al. 2012; R.G. Wallace and Kock 2012; Jones et al. 2013; Maye et al. 2014; Degeling et al. 2015; Mann et al. 2015; Wiethoelter et al. 2015; R. Wallace et al. 2016).

In this volume, we present the emergence of an urbanized Ebola in West Africa in late 2013, spreading from human to human and infecting 28,000, as a quintessential example of such a neoliberal transition. Societal shifts extending from local environmental and social spaces out to global relational geographies turned what until this point had been a backwater virus into a sudden protopandemic threat.

The first paper collected here, published in fall of 2014 as the outbreak in West Africa finally captured the globe's broader attention as possibly more than a regional threat, positions the outbreak as global in origins. R.G. Wallace et al. (2014) describe the likely cascade by which multilateral structural adjustment and a multinational land rush encroached upon regional forests and truncated medical infrastructure. The resulting increases in Ebola spillover likely accelerated the emergence of a human-to-human infection. The authors hang the specific mechanism on the spread of monoculture oil palm to which Ebola-bearing species of bats are attracted and along which the latter's interface with humans likely expanded (Shafie et al. 2011).

The team pegs such epidemiological shifts, of which Ebola's urbanization appears representative, to a model of changes in environmental stochasticity (R. Wallace and R.G. Wallace 2015). The ecosystemic fluctuations across populations of a "noisy" forest typically truncate the chains of transmission on which such a virulent pathogen depends. Such noise asymptotically drives the pathogen population to local extirpation. When monoculture production is suddenly imposed upon such an agroforestry, the inherent disruption to pathogen transmission is stripped out, accelerating spillover and explosive growth across the host populations that remain.

The second paper, published here for the first time, revisits an outbreak of another Ebola, the Reston species, among industrial hog in the Philippines in 2009 and more recently in China (Barrette et al. 2009; Pan et al. 2014). The shifts in stochasticity at the heart of the first paper are extended into a more realistic travel-based formalism of spatial spread, an agroecological logic gate for epidemic control, and a more explicit connection to economic models of agricultural production by way of a variation of the Black-Scholes approach to option pricing.

The latter model shows that pathogen emergence in intensive production may outpace the margins agrifood extends to biocontrol. The financial holes that result appear largely filled only after the outbreak begins and by other stakeholders entirely. Livestock, contract farmers, public health, smallholder production, wildlife, the environment, and governments across administrative units are asked to bear the costs in health and/or finances. Should such substantial costs be returned to company ledgers, agribusiness as we know it would cease to exist. R. Wallace et al. (2016) hypothesize that as industrial livestock expands, in contrast, including across areas of Africa in which Ebola has already emerged as a human infection, multiple novel phenotypes are likely to emerge out of such an agroeconomic frame.

The third paper expands the epistemological implications of the West African outbreak, particularly within the context of what appears to be the development of an efficacious vaccine (Henao-Restrepo et al. 2015). Blips in new cases continue to intermittently reappear across the three countries most affected—Guinea, Liberia, and Sierra Leone—even as the outbreak is repeatedly declared ended (e.g., Barbarossa et al. 2015). The paper hypothesizes that the social environment driving the ecotypic shift in Ebola in the region may be strong enough to offset the epidemiological advances offered by the new vaccine and other proximate interventions (R.G. Wallace et al. 2016). Emergency and palliative responses cannot be separated from structural context. The virus, even when maneuvered back upon its proverbial heels, threatens to reemerge by virtue of causes beyond the biomedical, however important the latter.

The fourth paper, published here for the first time, applies a control theory model that prioritizes social policy and economic structure. The ecology of infectious disease emerges from more than aggregate measures of population dynamics, depending decisively upon such context.

The fifth paper, again published here first, examines mechanisms by which a deadly pandemic may become the trigger for an avalanche of socioeconomic disintegration that can carry far heavier burdens of morbidity and mortality than the disease outbreak itself.

Indeed, big picture, these papers together explore the relationships among the ostensible objects driving or dragging upon the disease—the pathogen and vaccines against, for instance—and the fields within which such objects relate. As we review here, growing banks of data at the molecular and clinical levels indicate that the Ebola strain in West Africa is little different from its forest predecessors. As causality emerges from the interaction between objects and field, a shift in regional agroeconomics can engender a parallax effect in which an unchanged object expresses new properties by virtue of a shift in context alone (Zizek 2006).

In short, as humanity's epidemiological history attests, context is more than merely a veritable stage on which pathogens and immunity clash. The regional agroeconomic impacts of global neoliberalism are foundational, felt across biocultural organization, down so far as virion and molecule. Exploring such connections frames what will likely be a cutting-edge science of the twenty first century (R.G. Wallace et al. 2015).

Minneapolis, MN, USA   Robert G. Wallace

References

Barbarossa, M. V., Dénes, A., Kiss, G., Nakata, Y., Röst, G., & Vizi, Z. (2015). Transmission dynamics and final epidemic size of Ebola Virus Disease outbreaks with varying interventions. *PLoS One, 10*(7), e0131398. doi:10.1371/journal.pone.0131398.

Barrette, R. W. Metwally, S. A., Rowland, J. M., Xu, L., Zaki, S. R., Nichol, S. T., et al. (2009). Discovery of swine as a host for the Reston ebolavirus. *Science, 325*(5937), 204–206. doi:10.1126/science.1172705.

Colgrove, J. (2002). The McKeown thesis: A historical controversy and its enduring influence. *American Journal of Public Health, 92,* 725–729.

Centeno, M. A., & Cohen, J. N. (2012). The arc of neoliberalism. *Annual Review of Sociology, 38*, 317–340. doi:10.1146/annurev-soc-081309-150235.

Degeling, C., Johson, J., Kerridge, I., Wilson, A., Ward, M., Stewart, C., et al. (2015). Implementing a One Health approach to emerging infectious disease: Reflections on the socipolitical, ethical and legal dimensions. *BMC Public Health, 15*, 1307. doi:10.1186/s12889-015-2617-1.

Engering, A., Hogerwerf, L., & Slingenbergh, J. (2013). Pathogen host environment interplay and disease emergence. Emerging Microbes and Infections, 2, e5. http://www.nature.com/emi/journal/v2/n2/full/emi20135a.htm.

FAO (2013b). *World Livestock 2013: Changing disease landscapes*. United Nations, Rome: Food and Agriculture Organization.

Ganti, T. (2014). Neoliberalism. *Annual Review of Anthropology*, 43, 89–104. doi:10.1146/annurev-anthro-092412-155528.

Harvey, D. (2005). *A brief history of Neoliberalism*. Oxford: Oxford University Press.

Henao-Restrepo, A., Longini, I. M., Egger, M., Dean, N. E., Edmunds, W. J., Camacho, A., et al. (2015). Efficacy and effectiveness of an rVSV-vectored vaccine expressing Ebola surface glycoprotein: Interim results from the Guinea ring vaccination cluster-randomised trial. *Lancet*. pii: S0140-6736(15)61117-5. doi:10.1016/S0140-6736(15)61117-5.

Johnson, S. (2006). *The ghost map: The story of London's most terrifying epidemic–and how it changed science, cities, and the modern world*. New York: Riverhead Books.

Jones, B. A., Grace, D., Kock, R., Alonso, S., Rushton, J., Said, M. Y., et al. (2013). Zoonosis emergence linked to agricultural intensification and environmental change. *PNAS, 110*, 8399–8404.

Kock, R. A., Alders, R., & Wallace, R. G. (2012). Wildlife, wild food, food security and human society. In Animal health and biodiversity - preparing for the future. Illustrating contributions to public health (pp. 71e79). Compendium of the OIE Global Conference on Wildlife, 23–25 February 2011, Paris.

Mann, E., Streng, S., Bergeron, J., & Kicher, A. (2015). A review of the role of food and the food system in the transmission and spread of Ebolavirus. *PLoS Neglected Tropical Diseases, 9*, e0004160. doi:10.1371/jorunal.pntd.0004160.

Maye, D., Dibden, J., Higgens, V., & Potter, C. (2012). Governing biosecurity in a neoliberal world: Comparative perspectives from Australia and the United Kingdom. *Environment and Planning A, 44*, 150–168.

Maye, D., Enticott, G., Naylor, R., Ilbery, B., & Kirwan, J. (2014). Animal disease and narratives of nature: Farmers' reactions to the neoliberal governance of bovine Tuberculosis. *Journal of Rural Studies, 36*, 401–410.

McNeill, W. H. (1977/2010). *Plagues and peoples*. New York: Anchor Books.

Pan, Y., Zhang, W., Cui, L., Hua, X., Wang, M., & Zeng, Q. (2014). Reston virus in domestic pigs in China. *Archives of Virology, 159*(5), 1129–1132. doi:10.1007/s00705-012-1477-6. Epub 2012 Sep 21.

Pepin, J. (2011). The origins of AIDS. Cambridge: Cambridge University Press.

Plehwe, D., Walpen, B., & Neunhöffer, G. (Eds.). (2006). *Neoliberal hegemony: A global critique*. New York: Routledge.

Shafie, N. J., Sah, S. A. M., Latip, N. S. A., Azman, N. M., & Khairuddin, N. L. (2011). Diversity pattern of bats at two contrasting habitat types along Kerian River, Perak, Malaysia. *Tropical Life Sciences Research, 22*(2), 13–22.

Timberg, C., & Halpern, D. (2013). *Tinderbox: How the West Sparked the AIDS Epidemic and How the World Can Finally Overcome It*. New York: Penguin Press.

Wallace, R. G. (2010). King Leopold's pandemic. *Farming pathogens*. Available online at https://farmingpathogens.wordpress.com/2010/03/02/king-leopolds-pandemic/

Wallace, R., Bergmann, L., Hogerwerf, L., Kock, R., & Wallace, R. G. (2016, this volume). Ebola in the hog sector: Modeling pandemic emergence in commodity livestock.

Wallace, R., & Wallace, R. G. (2015). Blowback: New formal perspectives on agriculturally driven pathogen evolution and spread. *Epidemiology and Infection, 143*(10), 2068–2080.

Wallace, R. G., Bergmann, L., Kock, R., Gilbert, M., Hogerwerf, L., Wallace, R., et al. (2015). The dawn of structural one health: A new science tracking disease emergence along circuits of capital. *Social Science & Medicine, 129*, 68–77.

Wallace, R. G., & Kock, R. A. (2012). Whose food footprint? Capitalism, agriculture and the environment. *Human Geography, 5*(1), 63–83.

Wallace, R. G., Kock, R., Bergmann, L., Gilbert, M., Hogerwerf, L., Pittiglio, C., et al. (2016). Did neoliberalizing West African forests produce a new niche for Ebola? *International Journal of Health Services, 46*, 149–165.

Wallace, R. G., Gilbert, M., Wallace, R., Pittiglio, C., Mattioli, R., & Kock, R. (2014). Did Ebola emerge in West Africa by a policy-driven phase change in agroecology? *Environment and Planning A, 46*(11), 2533–2542.

Watts, S. (1997). *Epidemics and history: Disease, power, and imperialism.* New Haven: Yale University Press.

Wiethoelter, A., Beltran-Alcrudo, D., Kock, R., & Mor, S. (2015). Global trends in infectious dieases at the wildlife-livestock interface. *Proceedings of the National Academy of Sciences of the United States of America, 112*, 9962–9967.

Wolfe, N. D., Dunavan, C. P., & Diamond, J. (2007). Origins of major human infectious diseases. *Nature, 447*, 279–283.

Zizek, S. (2006). *The parallax view.* Cambridge, MA: MIT.

Contents

About the Authors

Luke Bergmann is an assistant professor in the Department of Geography at the University of Washington, USA. Much of his research explores how globalization affects human-environment relations through several thematic focuses. These include changes in the evolution and spread of pathogens, as well as studies of shifting socio-spatial relationships between carbon, land, capital, and consumption.

Marius Gilbert is an agricultural scientist based at the Université Libre de Bruxelles in Belgium. He has a strong interest in the spatial epidemiology of infectious diseases and the ways emerging diseases are influenced by the intensification of livestock production systems. Gilbert consults with the International Livestock Research Institute and the Food and Agriculture Organization on the Gridded Livestock of the World database, a unique spatial data set describing how livestock are distributed globally and how this distribution is transformed through agricultural intensification.

Lenny Hogerwerf is an epidemiologist and disease ecologist at the Centre for Infectious Disease Control (CIb) of the National Institute for Public Health and the Environment (RIVM) of the Netherlands. She obtained her PhD on the epidemiology of Q fever in dairy goat herds in the Netherlands at the Department of Farm Animal Health of the Faculty of Veterinary Medicine of Utrecht University. She has also developed a variety of ecological models for explaining the dynamics of HPAI H5N1 and disease landscapes more generally. She has consulted for the Food and Agriculture Organization and Veterinaires Sans Frontieres Belgium.

Richard Kock is a wildlife veterinarian, researcher, and conservationist. He is the chair in wildlife health and emerging diseases at the Pathology and Pathogen Biology Department at the Royal Veterinary College in London. For over three decades, he has participated in a variety of research and management efforts in epizoology, wildlife health, and conservation in countries around the world, including across sub-Saharan Africa. His most recent projects include the eradication of rinderpest and efforts to control foot-and-mouth disease and equine piroplasmosis in wild populations in a context of increasing agricultural intensification.

Raffaele Mattioli, DVM, PhD, is senior officer of disease ecology/noninfectious & production disease at the Food and Agriculture Organizations of the United Nations (FAO), Rome, Italy. He holds a PhD in tropical animal production at the Prince Leopold Institute of Tropical Medicine, Antwerp, Belgium. He has more than 16 years of working experience in Africa and more than 30 years of working experience with developing countries. In FAO Rome, his geographical regions of competence are the Middle East, Latin America and the Caribbean, and sub-Saharan Africa. Main activities range from livestock-agriculture policy development to integrating pest and animal health management. Mattioli is the author of more than 100 scientific publications in peer-reviewed scientific journals, book chapters, and international conference proceedings.

Claudia Pittiglio is an ecologist and a GIS expert in landscape ecology, species distribution modeling, risk mapping for animal diseases and human-wildlife conflicts, spatial analysis, and remote sensing. She holds a PhD in landscape ecology from the Faculty of Geo-Information Science and Earth Observation, University of Twente, in the Netherlands. Pittiglio currently works for the Animal Production and Health Division, Food and Agriculture Organization of the United Nations (FAO), in Rome, Italy. She has conducted modeling on crop raiding in elephants, poverty in Uganda, Rift Valley fever, and human-wildlife interfaces on sites across Africa.

About the Editors

Robert G. Wallace, PhD, is a public health phylogeographer presently visiting the University of Minnesota's Institute for Global Studies. His research has addressed the evolution and spread of influenza, the agroeconomics of Ebola, the social geography of HIV/AIDS in New York City, the emergence of Kaposi's sarcoma herpesvirus out of Ugandan prehistory, and the evolution of infection life history in response to antivirals. Wallace is co-author of *Farming Human Pathogens: Ecological Resilience and Evolutionary Process* (Springer) and author of *Big Farms Make Big Flu: Dispatches on Infectious Disease, Agribusiness, and the Nature of Science* (Monthly Review Press). He has consulted for the Food and Agriculture Organization of the United Nations and the Centers for Disease Control and Prevention.

Rodrick Wallace, PhD, is a research scientist in the Division of Epidemiology of the New York State Psychiatric Institute at Columbia University. He received undergraduate and graduate degrees in mathematics and physics from Columbia, worked for a decade as a public interest lobbyist, is a past recipient of an Investigator Award in Health Policy Research from the Robert Wood Johnson Foundation, and is the author of numerous books and papers on matters of public health and public order.

Chapter 1
Did Ebola Emerge in West Africa by a Policy-Driven Phase Change in Agroecology?

Robert G. Wallace, Marius Gilbert, Rodrick Wallace, Claudia Pittiglio, Raffaele Mattioli, and Richard Kock

1.1 Ebola's Social Context

The ongoing outbreak of human Ebola virus in West Africa, the largest and most extensive recorded, began in forest villages across four districts in southeastern Guinea as early as December 2013 (Baize et al. 2014; Nishiura and Chowell 2014). The epidemic subsequently spread across Guinea, Liberia, and Sierra Leone, including capital cities Conakry and Monrovia, before infiltrating Nigeria and Senegal. With infections at this writing newly diagnosed off-continent, the outbreak represents a significant enough threat that the World Health Organization declared it a Public Health Emergency of International Concern (Briand et al. 2014).

Bausch and Schwarz (2014) hypothesize that the virus initially spilled over in Guinea as a result of a combination of national economic and political impacts upon the first epicenter's forest community. Poverty drives forest encroachment, the infected present at inadequate medical facilities amplifying transmission, and

R.G. Wallace (✉)
Institute for Global Studies, University of Minnesota, Minneapolis, MN, USA
e-mail: rwallace24@gmail.com

M. Gilbert
Biological Control and Spatial Ecology, Universite Libre de Bruxelles, Brussels, Belgium

R. Wallace
Division of Epidemiology, The New York State Psychiatric Institute, New York, NY, USA

C. Pittiglio • R. Mattioli
Animal Production and Health Division, FAO, United Nations, Roma, Italy

R. Kock
Pathology and Pathogen Biology, The Royal Veterinary College, London, England

impoverished countries are buffeted by a cascade of logistical failures extending out from the outbreak itself to the very basics of societal function, including failing to provide even enough food.

The shifts in land use in Guinée forestière where the virus originated are also connected to the kinds of governmental policy promoting neoliberal structural adjustment that, alongside divesting public health infrastructure, open domestic food production to global circuits of capital (Moseley et al. 2010; World Bank 2014; Wallace et al. 2015). Under the newly democratized Guinean government, the British-backed, Nevada-based Farm Land of Guinea Limited, now of Africa, secured 99-year leases for two parcels totaling nearly 9000 hectares outside the villages of N'Dema and Konindou in Dabola prefecture, where a secondary epicenter developed, and 98,000 hectares outside the village of Saraya in Kouroussa prefecture (Farm Lands of Guinea 2011). The new acquisitions, a symbolic start, are to be developed for maize and soybean. The Ministry of Agriculture has tasked the company to survey and map an additional 1.5 million hectares for third-party development.

International deals represent the latest in a series of postcolonial efforts to increase agricultural production in Guinea, such as in rice and coffee, including Forested Guinea (Morris et al. 2009). Bausch and Schwarz (2014) characterize the area as a mosaic of small and isolated populations of a variety of ethnic groups that hold little political power and receive little social investment. Its economy and ecology are also strained by thousands of refugees from civil wars in neighboring countries. The forest is subject to the tandem trajectories of accelerating deterioration in public infrastructure and concerted efforts at private development dispossessing smallholdings and traditional foraging grounds for mining, clear-cut logging, and increasingly intensified agriculture.

Relationships across global production, deforestation, and development are rarely direct or deterministic, however (Bergmann 2013). In-country, complex combinatorials of ownership, lobbying, popular protest, and governance foreign and domestic imprint upon local outcomes (Mendick 2013). Twice in 2007-2008, when commodity speculation drove up food prices worldwide, Guineans at large, objecting to shortages and high prices, organized general strikes that forced the military government at the time to block all exports, including agricultural, forestry, livestock, fisheries, and petroleum products (Berazneva and Lee 2013).

1.2 Palm Oil Hypothesis

Historically, new epidemiologies shape and are shaped by such contingencies (Watts 1997). By one hypothesis Ebola's latest spillover took place by way of a phase change in the local agroeconomics of oil palm *Elaeis guineensis*.

Natural and semiwild groves of different oil palm types *dura*, *pisifera*, and *tenera* have long served as a source of red palm oil in Forested Guinea (Madelaine et al. 2008; Delarue and Cochet 2013). Contrary to prelapsarian fantasies of hunting/gathering, forest farmers have been cultivating oil palm in one form or another for hundreds of years. Fallow periods, however, were reduced over the twentieth century from 20 years in the 1930s to ten by the 1970s, and still further by the 2000s, with the added effect of increasing grove density.

Other crops are grown in the forest too, of course (Fairhead and Leach 1996; Madelaine et al. 2008). Regional shade agriculture includes coffee, cocoa, and kola. Slash-and-burn rice, maize, hibiscus, and corms of the first year, followed by peanut and cassava of the second, and a fallow period, are rotated through the agroforest. Lowland flooding supports rice. Rotation carries cultural water other than functional crop succession, also including food security, land tenure, labor availability, and regional price fluctuations (Delarue and Cochet 2013).

The key point is that, in spite of increasing intensification, production sans finance capital can still be classified here as agroforestry.

Yet, as ever, the forest is again changing. With two million hectares of natural and traditionally cultivated stands, Guinea, trailing even Liberia's weak sector, recently began a push toward commoditizing oil palm in the face of cheap imports from Asia (Carrere 2010; Ferrand et al. 2012). As of 2007, government plans included expanding family and industrial production to 15,000 hectares and 84,000 tons of palm oil by 2015, more than half of the latter to be produced by Forested Guinea plantations (Carrere 2010).

The Guinean Oil Palm and Rubber Company (SOGUIPAH) founded by the state in 1987 began along the lines of a parastatal cooperative, since developing into a full-fledged state company (Delarue and Cochet 2013). SOGUIPAH is leading regional efforts that began in 2006 to develop plantations of intensive hybrid palm for commodity export. The company has economized palm production in Yomou prefecture, south of the outbreak area, by contract farming, negotiating land requisition, organizing supply chains, franchising a production model, and, backed by police, expropriating farmland, the latter intermittently setting off violent protest. In 2011 villagers were run off their rice, coffee, and rubber fields, and forced to find refuge in a church in Nzarakora, the provincial capital (AFP 2011).

International aid has accelerated the transition. An industrial palm mill financed by the European Investment Bank permitted SOGUIPAH four times the capacity of its previous mill (Carrere 2010). The new mill ended the artisanal extraction that as late as 2010 provided local populations with full employment. The subsequent increase in seasonal production has at one and the same time led to harvesting above the mill's capacity and operation below capacity off-season, leading to a conflict between the company and 2000 of its now partially proletarianized producers and pickers, some of whom insist on processing a portion of their own yield to cover the resulting gaps in cash flow. Contractors who insist on processing their own oil during the rainy season now risk arrest.

Discipline is also imposed by the biology of the industrialized tree. Producers who break their contracts are left with F2 that produce only 60 % the oil of first-generation seeds provisioned by the company (Delarue and Cochet 2013).

The new economic geography also instantiates a classic case of land enclosure, turning a tradition of shared forest commons toward expectations that informal pickers working fallow land outside their family lineage will obtain an owner's permission before picking palm (Madelaine 2005; Carrere 2010). Concurrently, some of the smallholders who have remained independent have adapted to the new environment. Farmers surveyed around the village of Nienh, south of the initial outbreak, preferred planting hybrid palm in monoculture slash-and-burn for increased oil production and income, and for the private control of a resource and the land under it (Madelaine et al. 2008). SOGUIPAH has had additional effects on production in areas where it claims no plantations or contractors, including technology transfer and upon commodity prices (Delarue and Cochet 2013).

The agroeconomic shifts in all their complexities appear to interpenetrate the forest's epizoology.

The earliest documented cases of the outbreak in West Africa appear to be a 2-year-old village boy and his 3-year-old sister north of Guéckédou, a town of 200,000. But the focus on an index case "patient zero" may miss the point. Ebola may have been circulating for years, something Hewlett and Amola's (2003) work in Uganda suggests local populations may even have recognized. Indeed, Schoepp et al. (2014) found antibodies to multiple species of Ebola, particularly the Zaire strain, in patients in Sierra Leone as far back as 5 years ago. Phylogenetic analyses of the virus's own genome date the lower bound of Zaire Ebola's entry into West Africa a decade ago (Dudas and Rambaut 2014; Gire et al. 2014).

Figure 1.1(a) shows an archipelago of oil palm plots in the Guéckédou area, the outbreak's apparent ground zero. Land use appears as a mosaic of local villages surrounded by dense vegetation interspersed with fruit-tree plantations (Figure 1.1b–d), an environment also suitable for frugivore bats (*Pteropodidae*), a key Ebola reservoir, and other small crop-raiding species. Shafie et al. (2011) document a variety of disturbance-associated fruit bats attracted to oil palm plantations. Bats migrate to oil palm for food and shelter from the heat while the plantations' wide trails permit easy movement between roosting and foraging sites. Oil palm picking in the area occurs year-round, but the biggest push takes place at the start of the dry season when multiple Ebola outbreaks have begun across the Sub-Sahara (Carrere 2010; Bausch and Schwarz 2014).

Even in West Africa, the western edge of the traditional range of *Pteropodidae*, fruit bats are likely to display the kind of plastic biogeography migratory waterfowl have demonstrated feeding on waste grain hundred kilometers north of their destroyed natural habitat (Cooke et al. 1995). Leroy et al. (2009) tracked Ebola transmission in the Democratic Republic of Congo to massive hunting during the annual fruit bat migration up the Lulua River, including the hammer-headed bat (*Hypsignathus monstrosus*) and Franquet's epauletted fruit bat (*Epomops franqueti*), two of the three species likely to be Ebola reservoirs. Bats occupied the outbreak area for several weeks, roosting in fruit trees and the palm trees of a massive abandoned plantation bats had been visiting for half a century.

Bushmeat need not be a default explanation for any given outbreak, however. Field (2009) noted that deforestation, including from oil palm planting, changes foraging behavior of the flying fox, fixating now on horticulture crops, and expands

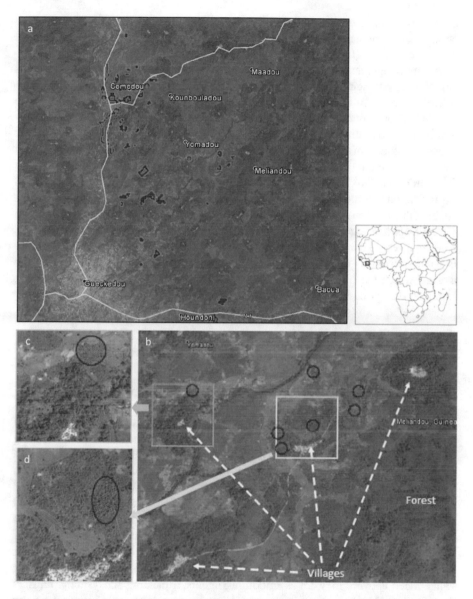

Fig. 1.1 Land-use pattern in West African Ebola's putative area of origin near Guéckédou, Guinea. The characteristic landscape is a mosaic of villages surrounded by dense vegetation and interspersed by crop fields of oil palm (*red*) and patches of open forest and regenerated young forest. The general pattern can be discerned at a coarse spatial scale north of Guéckédou (**a**) and a finer scale west of Meliandou (**b–d**).

interfaces among bats, humans, and livestock. Fruit bats in Bangladesh transmitted Nipah virus to human hosts by urinating on the date fruit of the planted palm trees humans cultivated (Luby et al. 2009).

A similar agroecology characterizes the Kailahun epicenter in Liberia. Guinea's neighbor to the south, however, hosts a different trajectory in agricultural consolidation, back to Firestone Rubber Company's first investments in 1925 and a post-WWII open door policy privatizing land across sectors, including rubber, timber, iron-ore, and diamonds (Fouladbash 2013). More recently, alongside a longer national tradition of wage labor, international logging, mining, and agro-industrial companies, including palm oil companies Sime Darby (Malaysia), Equatorial Palm Oil (UK), and Golden Veroleum (Indonesia), have partaken in large-scale land expropriation totaling a third of the country's land surface, with planned concessions pushing that total to nearly 45 % (Evans and Griffiths 2013; Murombedzi 2014).

Development schemata look to regionalize that model (Zagema 2011). The hot zone as a whole comprises a part of the larger Guinea Savannah Zone the World Bank describes as "one of the largest underused agricultural land reserves in the world," which the Bank sees best developed by market commercialization, if not solely on the agribusiness model (Morris et al. 2009).

1.3 A General Model of Epidemiological Phase Change

The functional relationships that such economic relationships share with epidemiology have been modeled elsewhere.

R. Wallace and R.G. Wallace (2014) describe the sudden emergence of poliomyelitis in postwar England and Wales as the consequence of the increasing per capita affluence of its human host. The normalized extents of polio epidemics between 1940 and 1957 are a function of the millions of registered cars in the UK. The sudden availability of private travel in the UK after World War II appeared sufficient to breach the level of isolation between small focal polio outbreaks that kept the larger island "sterilized." The phase transition of its host's sociogeographic space triggered polio's punctuated emergence as an epidemic.

The normalized characteristic extent of an epidemic can be viewed as a classic order parameter. For polio the number of cars or for Ebola palm oil's expanding value-added network may act as smoothing agents, truncating ecosystemic barriers that interrupt chains of transmission, and driving a coevolutionary socioviral system across a critical point.

Taking the perspective of R. Wallace and R.G. Wallace (2014), the most direct mathematical approach involves a stochastic extinction model of an exponentially growing population.

Let $N_t \geq 0$ represent the number of individuals of a particular pathogen "species" at time t. The simplest possible model is given by the stochastic differential equation

$$dN_t = \alpha N_t dt + \sigma N_t dW_t^H \tag{1.1}$$

where $\alpha > 0$ is a characteristic rate constant for exponential growth, σ is an index of "noise" strength, and dW_t^H represents a fractional white noise process with index $0 < H < 1$. $H = 1/2$ represents ordinary white noise. Fractional white noise is defined by the covariance relation

$$\text{Cov}(W^H(t), W^H(s)) = 1/2(t^{2H} + s^{2H} - |t - s|^{2H}) \tag{1.2}$$

Taking the Ito formula expansion for $\log(N_t)$, the correction factor from the added noise term gives the classic result (Zeng et al. 2013)

$$N_t = N_0 \exp\left[\alpha t - \frac{\sigma^2}{2}t^{2H} + \sigma W_t^H\right] \tag{1.3}$$

Following Zeng et al. (2013) Theorem 3.1, if $0 < H < 1/2$, then the system is asymptotically explosively unstable for any $\alpha > 0$. For $H = 1/2$, if $\sigma^2 > 2\alpha$, noise-driven fluctuations will asymptotically drive the pathogen to the absorbing state of local extinction: that is, $N_t \to 0$. However, for noise defined by $1/2 < H < 1$, the system is always asymptotically stable in this sense, regardless of α.

Figure 1.2 shows two simulation examples for $H = 1/2$, with σ below and above criticality. For the first, below the critical index, the initial infection explodes by exponential growth. For the second, just above criticality, the initial infection eventually collapses toward extirpation.

A similar stochastic differential equation approach has been used to model noise-driven criticality in physical systems (Van den Broeck et al. 1997), suggesting that a more conventional phase transition methodology might provide particular insight, as done in R. Wallace and R.G. Wallace (2014).

A simple spatial analysis leads to similar results. Assuming a one-dimensional diffusion-growth model,

$$\partial N/\partial t = \mu \partial^2 N/\partial x^2 + \alpha N \tag{1.4}$$

where x represents distance, μ is the diffusion coefficient, and the initial patch size has length L, a Fourier series expansion leads to an exponential term in time (Okubo 1980),

$$\exp[(\alpha - \mu\pi^2/L^2)t] \tag{1.5}$$

Thus the infection dies out if $L < L_c \equiv \pi\sqrt{\mu/\alpha}$. In two dimensions, π is replaced by $4.81\dots$. The circumference/area ratio grows $\propto 1/L$, so that, for small L, edge habitat, rather than patch habitat, dominates disease ecology. In general, edge habitats, along which pathogens are likely better able to cross into humans, will not *incubate* pathogen populations as productively as patches. Indeed, a simple dimensional analysis leads to the $L_c \propto \sqrt{\mu/\alpha}$ expression. Expansion of Eq. (1.4) to a full stochastic differential equation model, with fractional time diffusion and colored spatial diffusion, is mathematically nontrivial (Balan and Tudor 2008).

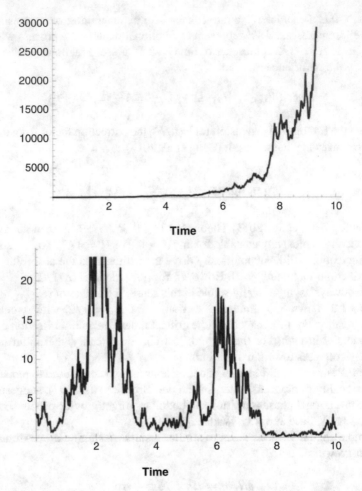

Fig. 1.2 Simulations of Eq. (1.1) using the white noise ($H = 1/2$) ItoProcess construct in Mathematica: $\alpha = 1$, critical $\sigma = \sqrt{2}$, initial number of infecteds, $N_0 = 3$. The *upper trace* has $\sigma = 0.5$, the lower has $\sigma = 1.6$. One thousand time steps. While the *upper outbreak* eventually reaches tens of thousands, the *lower* remains under 100.

The generality of the model opens any number of ways of operationalizing it. Again following R. Wallace and R.G. Wallace (2014), one might well populate both σ and H with variables beyond the stalwarts of population densities and transmission rates, instead including governmental policies promoting neoliberal or neocolonial development, an expansion in logging, or the elimination of traditional systems of subsistence that *at the population level* previously isolated pathogens or their carriers from livestock and/or humans, even as individual spillovers may have accrued with regularity.

The details are likely to differ, of course, but we hypothesize that such large-scale shifts across anthropogenic systems will routinely trigger sudden spikes in pathogen population load. The resulting increases in circulating pathogen and their phylogenesis may produce new patterns of infection and the evolution of increased transmissibility and virulence standard niche analyses by definition miss (e.g., Pigott et al. 2014). That is, changes in policy or socioeconomic structure, including the economics driving plantation farming, can "desterilize" a natural or human ecosystem in which a pathogen has been largely held in check at a low-level equilibrium value, or simply had not previously evolved.

In the other direction, traditional or conservation agricultures can by their diversity in time, space, and mode create numerous functional and physical barriers—counterintuitively, a kind of ecosystemic noise comparable to a steril-izing temperature in a physical system—limiting in a case-by-case basis many a pathogen's evolution and spread (R.G. Wallace and Kock 2012; R. Wallace and R.G. Wallace 2014).

1.4   Conclusions, Caveats, and Next Steps

The deductive model introduced here suggests that a critical noise of environmental stochasticity can be empirically defined for each ecosystem, incorporating measures of the disease-specific barriers imposed upon pathogen evolution and propagation.

Alongside commoditizing fruit trees and the social effects of governmental policies favoring dispossession—rejuxtaposing populations of people and animal alike—by an epizoological Allee effect clear-cutting Forested Guinea may have lowered the ecosystemic "temperature" below which Ebola can be "sterilized" and controlled (Stephens et al. 1999; Hogerwerf et al. 2010). Indeed, the model suggests the possibility in some parameter spaces that a threshold can be lowered to such a point that no emergency intervention can drive the pathogen population low enough to burn out on its own, refuting the false dichotomy between structural and emergency interventions (e.g., Osterholm 2014).

Whether such a conundrum defines West Africa remains to be ascertained. The specifics are likely to be complex. Guinea's new agriculture is more nuanced than "industrial" against "smallholder" (Madelaine et al. 2008; Carrere 2010; Delarue and Cochet 2013). As elsewhere, including Thailand and Mexico, smallholders, farm cooperatives, and even state companies are faced with a choice as global markets shift and tariffs are removed on multinationals beginning to buy up domestic land (Wise 2010; Moran 2011; Walker et al. 2012). Either sell off or consolidate, scaling up to meet the competition. Even as there are presently no multinational plantations in Guinea, oil palm there represents a classic case of creeping consolidation, enclosure, commoditization, and capitalization curtailing artisanal production. So while no private companies presently plants oil palm in Guinea, by a relational geography the effects of the global market upon the local agroecology appear to be felt already.

Other explanations for Ebola in West Africa are, of course, in play. Deforestation, dedevelopment, population mobility, periurbanization, cycle migration, and an inadequate health system that failed to recognize and isolate cases may have synergistically eased the ecosystemic friction acting on circulating Ebola. The lethargy of the international response exacerbated matters once the outbreak began. A Structural One Health may be able to unify such a variety of deterministic sources under the rubric Ebola's agroecology and its failure of containment—only 900 beds are presently available across all of West Africa—arose together out of the neoliberal program (and a longer history of exploitation) (Jones 2011; Wallace et al. 2015). Successful intervention for a pathogen circulating in such a system may call for rolling back the structural violence long visited upon the region.

The model here, itself part of the hypothesis, generalizes relationships among a (pathogen) population, its carry capacity, and stochastic noise (from a variety of possible sources) on punctuated dynamics. Ostensibly the model could be conditionalized in various ways for testing in specific systems, Ebola in West Africa included, but its simplicity here speaks to a general condition across systems, regardless of their biologies. Namely, pathogen success is fundamentally integrated with its population biological (and sociological) context.

Be that as it may, the statistical economic geographies, value-chain analyses, and dynamic socioecological niche modeling needed to test the hypothesis are welcome. Science proceeds, even as we must also recognize its difficulties. The data to test the hypothesis—e.g., remote sensing of owner-coded plantations across the region, geocoded prevalences of Ebola in symptom-free bats, and surveys of local epidemiological knowledge—are presently unavailable (Leach 2014). As the present outbreak signals, however, such efforts are critical for characterizing the ecosystems on which humanity must routinely be reminded it depends.

References

AFP (2011, August 3). Scores displaced in Guinea land grab row. Available online at http://www.news24.com/Africa/News/Scores-displaced-in-Guinea-land-grab-row-20110803.

Baize, S., Pannetier, D., Oestereich, L., Rieger, T., Koivogui, L., Magassouba, N., et al. (2014). Emergence of Zaire ebola virus disease in guinea. *The New England Journal of Medicine.* doi:10.1056/NEJMoa1404505.

Balan, R., & Tudor, C. (2008). The stochastic heat equation with fractional-colored noise: Existence of the solution. *Alea, 4*, 57–87.

Bausch, D., & Schwarz, L. (2014). Outbreak of Ebola virus disease in Guinea: Where ecology meets economy. *PLOS Neglected Tropical Diseases, 8*, e3056.

Berazneva, J., & Lee, D. R. (2013). Explaining the African food riots of 2007-2008: An empirical analysis. *Food Policy, 39*, 28–39.

Bergmann, L. (2013). Bound by chains of carbon: Ecological-economic geographies of globalization. *Annals of the Association of American Geographers, 103*(6), 1348–1370.

Briand, B., Bertherat, E., Cox, P., Formenty, P., Kieny, M. P., Myhre, J. K., et al. (2014). The international Ebola emergency. *The New England Journal of Medicine.* doi:10.1056/NEJMp1409858.

Carrere, R. (2010). *Oil palm in Africa: Past, present and future scenarios.* Montevideo: World Rainforest Movement.

Cooke, F., Rockwell, R. F., & Lank, D. B. (1995). *The snow geese of La Perouse Bay: Natural selection in the wild.* Oxford: Oxford University Press.

Delarue, J., & Cochet, H. (2013). Systemic impact evaluation: A methodology for complex agricultural development projects. The case of a contract farming project in Guinea. *European Journal of Development Research, 25*(5), 778–796.

Dudas, G., & Rambaut, A. (2014). Phylogenetic analysis of Guinea 2014 EBOV ebolavirus outbreak. *PLoS Currents Outbreaks* (1st ed.). doi:10.1371/currents.outbreaks.84eefe5ce43ec9dc0bf0670f7b8b417d.

Evans, R., & Griffiths, G. (2013). *Palm Oil, land rights and ecosystem services in Gbarpolu County, Liberia.* Research Note 3, Walker Institute for Climate System Research, University of Reading. Available online at http://centaur.reading.ac.uk/33817/1/WalkerInResNote3.pdf.

Fairhead, J., & Leach, M. (1996). *Misreading the African landscape: Society and ecology in a forest-savanna mosaic* (374pp.). London: Cambridge University Press.

Farm Lands of Guinea (2011). Farm Lands of Guinea completes reverse merger and investment valuing the company at USD 45 million. PR Newswire. Available online at http://www.bloomberg.com/apps/news?pid=newsarchive&sid=a9cwc86wQ3zQ.

Ferrand, P., Koundiuno, J., Thouillot, F., & Camara, K. (2012). Enjeux de la filiare huile de palme en Rapublique de Guinae. *Grain de Sel, 58*, 36–38.

Field, H. E. (2009). Bats and emerging zoonoses: Henipaviruses and SARS. *Zoonoses and Public Health, 56*, 278–284. doi:10.1111/j.1863-2378.2008.01218.x.

Fouladbash, L. (2013). *Agroforestry and shifting cultivation in Liberia: Livelihood impacts, carbon tradeoffs, and socio-political obstacles.* PhD thesis, Natural Resources and Environment, University of Michigan. Available online at http://deepblue.lib.umich.edu/bitstream/handle/2027.42/100128/Lisa%20Fouladabash%20thesis%20Fall%2013.pdf?sequence=1.

Gire, S. K., Goba, A., & Andersen, K. G. (2014). Genomic surveillance elucidates Ebola virus origin and transmission during the 2014 outbreak. *Science, 345*, 1369–1372.

Hewlett, B. S., & Amola, K. G. (2003). Cultural contexts of ebola in Northern Uganda. *Emerging Infectious Diseases, 9*(10), 1242–1248.

Hogerwerf, L., Houben, R., Hall, K., Gilbert, M., Slingenbergh, J., & Wallace, R. G. (2010). *Agroecological resilience and protopandemic influenza.* Rome: Animal Health and Production Division, Food and Agriculture Organization.

Jones, J. (2011). Ebola, emerging: The limitations of culturalist discourses in epidemiology. *The Journal of Global Health, 1*, 1–6. Available online at http://www.ghjournal.org/?p=6194.

Leach, M. (2014, April 3). Ebola in Guinea; people, patterns, and puzzles. *The Lancet Global Health Blog.* Available online at http://globalhealth.thelancet.com/2014/04/03/ebola-guinea-people-patterns-and-puzzles.

Leroy, E. M., Epelboin, A., Mondonge, V., Pourrut, X., Gonzalez, J.-P., Muyembe-Tamfum, J.-J., et al. (2009). Human Ebola outbreak resulting from direct exposure to fruit bats in Luebo. Democratic Republic of Congo, 2007. *Vector-Borne and Zoonotic Diseases, 9*(6), 723–728. doi:10.1089/vbz.2008.0167.

Luby, S. P., Gurley, E. S., & Hossain, M. J. (2009). Transmission of human infection with Nipah Virus. *Clinical Infectious Diseases, 49*(11), 1743–1748.

Madelaine, C. (2005). *Analyse du fonctionnement et de la dynamique de la palmeraie sub-spontane en Guinee forestiere. Cas du village de Nienh.* (80pp.) MSc thesis, Agro. MAENGREF, Montpellier.

Madelaine, C., Malezieux, E., Sibelet, N., & Manlay, R. J. (2008). Semi-wild palm groves reveal agricultural change in the forest region of Guinea. *Agroforestry Systems, 73*, 189–204. doi:10.1007/s10457-008-9146-1.

Mendick, R. (2013, December 15). Tony Blair and the Africa mine deal. *The Telegraph.* Available online at http://www.telegraph.co.uk/news/politics/tony-blair/10518355/Tony-Blair-and-the-Africa-mine-deal.html.

Moran, L. S. (2011). *Oil palm plantations: Threats and opportunities for tropical ecosystems*. UNEP Global Environmental Alert Service. Available online at http://na.unep.net/geas/getUNEPPageWithArticleIDScript.php?article_id=73.

Morris, M. L., Binswanger-Mikhize, H. P., & Byerlee, D. (2009). *Awakening Africa's sleeping giant: Prospects for commercial agriculture in the guinea savannah zone and beyond*. Washington, DC: World Bank Publications.

Moseley, W. G., Carney, J., & Becker, L. (2010). Neoliberal policy, rural livelihoods, and urban food security in West Africa: A comparative study of The Gambia, Cote d'Ivoire, and Mali. *PNAS, 107*, 5574–5579.

Murombedzi, J. (2014). National and transnational land grabs in Africa: Implications for local resource governance. In G. Branes & B. Child (Eds.), *Adaptive cross-scalar governance of natural resources* (pp. 75–102). New York: Routledge.

Nishiura, H., & Chowell, G. (2014). Early transmission dynamics of Ebola virus disease (EVD), West Africa, March to August 2014. *Euro Surveillance, 19*(36), pii=20894.

Okubo, A. (1980). *Diffusion and ecological problems: Mathematical models*. Biomathematics (Vol. 10). New York: Springer.

Osterholm, M. T. (2014, August 1). What we need to fight Ebola. *Washington Post*. Available online at http://www.washingtonpost.com/opinions/what-we-need-to-fight-ebola/2014/08/01/41f4dbb8-182d-11e4-9349-84d4a85be981_story.html.

Pigott, D. M., Golding, N., Mylne, A., Huang, Z., Henry, A. J., Weiss, D. J., et al. (2014). Mapping the zoonotic niche of Ebola virus disease in Africa. *eLife, 3*, e04395.

Schoepp, R. J., Rossi, C. A., Khan, S. H., Goba, A., & Fair, J. N. (2014). Undiagnosed acute viral febrile illnesses, Sierra Leone. *Emerging Infectious Diseases, 20*(7), 1176–1182.

Shafie, N. J., Sah, S. A. M., Latip, N. S. A., Azman, N. M., & Khairuddin, N. L. (2011). Diversity pattern of bats at two contrasting habitat types along Kerian River, Perak, Malaysia. *Tropical Life Sciences Research, 22*(2), 13–22.

Stephens, P. A., Sutherland, W. J., & Freckleton, R. P. (1999). What is the Allee Effect? *Oikos, 87*, 185–190.

Van den Broeck, C., Parrondo, J., & Toral, R. (1997). Noise-induced nonequilibrium phase transition. *Physical Review E, 73*, 4084–4094.

Walker, P., Cauchemez, S., Hartemink, N., Tiensin, T., & Ghani, A. C. (2012). Outbreaks of H5N1 in poultry in Thailand: The relative role of poultry production types in sustaining transmission and the impact of active surveillance in control. *Journal of the Royal Society, Interface*. http://dx.doi.org/10.1098/rsif.2012.0022.

Wallace, R. G., Bergmann, L., Kock, R., Gilbert, M., Hogerwerf, L., Wallace, R., et al. (2015). The dawn of structural one health: A new science tracking disease emergence along circuits of capital. *Social Science & Medicine, 129*, 68–77.

Wallace, R. G., & Kock, R. A. (2012). Whose food footprint? Capitalism, agriculture and the environment. *Human Geography, 5*(1), 63–83.

Wallace, R., & Wallace, R. G. (2014). Blowback: New formal perspectives on agriculturally driven pathogen evolution and spread. *Epidemiology and Infection*. doi:10.1017/S0950268814000077.

Watts, S. (1997). *Epidemics and history: Disease, power, and imperialism*. New Haven: Yale University Press.

Wise, T. A. (2010). *Agricultural dumping under NAFTA: Estimating the costs of U.S. agricultural policies to Mexican producers*. Mexican Rural Development Research Report No. 7, Woodrow Wilson International Center for Scholars.

World Bank (2014). *Guinea - Agriculture Sector Support Project*. World Bank Group, Washington, DC. Available online at http://documents.worldbank.org/curated/en/2014/09/20204824/guinea-agriculture-sector-support-project.

Zagema, B. (2011). *Land and power: The growing scandal surrounding the new wave of investments in land*. Oxfam Briefing Paper 151, Oxford: Oxfam International.

Zeng, Z., Quan, C. Y., & Yang, Q. (2013). Almost sure and moment stability properties of fractional order Black-Scholes model. *Fractional Calculus and Applied Analysis, 16*, 317–331.

Chapter 2
Ebola in the Hog Sector: Modeling Pandemic Emergence in Commodity Livestock

Rodrick Wallace, Luke Bergmann, Lenny Hogerwerf, Richard Kock, and Robert G. Wallace

2.1 Introduction

Human impact is increasingly transforming planet Earth into planet Farm. Forty percent of the planet's surface is dedicated to agriculture, with many millions more hectares to be brought into production by 2050 (Foley et al. 2005, Alexandratos and Bruinsma 2012, FAO 2013a). Livestock, representing 72 % of global animal biomass, are simultaneously highly concentrated and widely dispersed across the planet's surface (Smil 2002; Van Boeckel 2013; Robinson et al. 2014) (Fig. 2.1). The livestock sector uses a third of available freshwater and a third of cropland for feed (Steinfeld et al. 2006; Herrero et al. 2013). Feed production, enteric fermentation, manure, animal processing, and transportation in turn produce greenhouse gases at 7.1 gigatonnes CO_2-eq per year (Gerber et al. 2013).

Agricultural impact extends to emergent disease. If by its global expansion alone, commodity agriculture increasingly acts as a nexus through which pathogens of diverse origins migrate from even the most isolated reservoirs in the wild to the most globalized of population centers (Graham et al. 2008; R.G. Wallace 2009;

R. Wallace
Division of Epidemiology, The New York State Psychiatric Institute, New York, NY, USA

L. Bergmann
Department of Geography, University of Washington, Seattle, WA, USA

L. Hogerwerf
Centre for Infectious Disease Control, National Institute for Public Health
and the Environment, Bilthoven, The Netherlands

R. Kock
Pathology & Pathogen Biology, The Royal Veterinary College, London, UK

R.G. Wallace (✉)
Institute for Global Studies, University of Minnesota, Minneapolis, MN, USA
e-mail: rwallace24@gmail.com

© Springer International Publishing Switzerland 2016
R.G. Wallace, R. Wallace (eds.), *Neoliberal Ebola*,
DOI 10.1007/978-3-319-40940-5_2

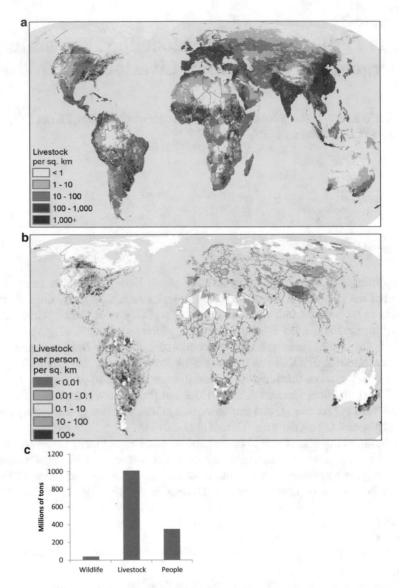

Fig. 2.1 Global livestock. (**a**) Total livestock (cattle, chickens, ducks, pigs, sheep, and goats) per km[2] (2006) (Eckert IV projection). Sixty-four percent of all cattle, chickens, ducks, pigs, sheep, and goats are found on 2 % of Earth's land surface. At the same time, 10 % of these stocks are found across 69 % of land surface. Data from the Gridded Livestock of the World v2.0 (Robinson et al. 2014). (**b**) Livestock per human per km[2]. Livestock data from Gridded Livestock of the World v2.0 (Robinson et al. 2014). Human data from Global Rural–Urban Mapping Project (GRUMP) v1.0 (2000) (Balk et al. 2006). All areas with less than 0.1 persons/km[2] masked out. (**c**) Global biomass wildlife, livestock, and people. UNFAO data.

Jones et al. 2013; Liverani et al. 2013; Engering et al. 2013; FAO et al. 2013b). The lengthier the associated supply chains and the greater the extent of adjunct deforestation, the more diverse (and exotic) the zoonotic pathogens that enter the food chain (Wolfe et al. 2005; R.G. Wallace et al. 2010; FAO et al. 2013b). Among such emergent pathogens are industrial *Campylobacter*, Nipah virus, Q fever, hepatitis E, *Salmonella enteritidis*, foot-and-mouth disease, and a variety of novel influenza variants, including H1N1 (2009), H1N2v, H3N2v, H5N1, H5N2, H6N1, H7N1, H7N3, H7N7, H7N9, and H9N2 (Epstein et al. 2006; Myers et al. 2006; Graham et al. 2008; Leibler et al. 2009; Jones et al. 2013; Khan et al. 2013).

Intensive agriculture's diseconomies of scale extend beyond the unintended epidemiological consequences of globalizing transport and distribution. Its production cycles degrade ecosystemic resilience to disease; complicate interventions by treating humans and animals as markets and commodities first; and by its genetic monocultures, high population densities, rapid throughput, and expanding exports accelerate pathogen spread and evolution (R.G. Wallace 2009; Mennerat et al. 2010; Drew 2011; Van Boeckel et al. 2012; R.G. Wallace and Kock 2012; Ercsey-Ravasz et al. 2012; Liverani et al. 2013; Khan et al. 2013; FAO 2013b; R.G. Wallace et al. 2015).

The recent outbreak of *Zaire ebolavirus* in West Africa appears another case of a pathogen's agroeconomic transition. ZEBOV, among other Ebola species, had been circulating in the area for at least a decade before emerging as a human-to-human infection (Schoepp et al. 2014; Dudas and Rambaut 2014; Gire et al. 2014). Sudden Ebola spillovers in the region were already associated with socioeconomically driven shifts in land use. Formenty et al. (1999), for instance, described the outbreak of *Taï Forest ebolavirus* in Cote d'Ivoire as arising out of a combination of impacts upon Parc National de Taï:

> The habitat in the region has also been modified constantly by human migration from regions north of the forest belt. This process has sharply increased in the last 6 years since the start of the civil war in Liberia; the influx of refugees doubled the local population between early 1992 and early 1993 and again in the summer of 1994. Massive migration of humans with their domestic animals and other commensal organisms could explain environmental perturbation near the forest and consequently in the forest. The increased deforestation pressure resulting from this influx would also result in perturbation of the habitat. Crop activities have developed on the edge of the park and in the park itself. Illegal plantations and poaching into the Taï National Park have increased from 1985 to 1995 and led to the existence of a large area of farmland and broken forest. This area was only 2 km from the home range of the [possibly infected] chimpanzees that were studied.

R.G. Wallace et al. (2014) proposed ZEBOV emerged as a human-to-human infection late 2013 out of another such phase change in agroecology, in this case brought about by regional neoliberal development, juxtaposing multinational mining, logging, intensifying agriculture, and a structurally adjusted medical infrastructure (Bausch and Schwarz 2014; Kentikelenis et al. 2014). Neoliberalism is a program of political economy organized around globalizing laissez-faire economic liberalism, promoting free trade, and reducing state expenditures for the greater population in favor of protecting private property and deregulating economic markets (Harvey 2005; Centeno and Cohen 2012; Ganti 2014).

As a first attempt in framing the mechanism by which ZEBOV emerged in the region, R.G. Wallace et al. (2014) hypothesized the strain arose as oil palm, to which Ebola-bearing fruit bats are attracted, underwent a classic case of creeping consolidation, enclosure, commoditization, and proletarianization in the Guinée forestière. At one and the same time, the transition in agroforestry curtailed artisanal production and may have expanded the human–bat interface over which the virus crosses.

Saéz et al. (2015) have since proposed the initial ZEBOV spillover occurred when children outside Meliandou, Guinea, including the putative index case, caught and played with bats of an insectivore species (*Mops condylurus*) previously documented an EBOV carrier. Whatever the specific source, one of many in the region, shifts in agroeconomic context appear a primary causal factor. Previous studies show *Mops* also attracted to expanding cash crop production in West Africa, including sugar cane, cotton, and macadamia (Noer et al. 2012; Taylor et al. 2013; Stechert et al. 2014).

Indeed, in contrast to bushmeat, burial practices, and specific host reservoirs of arguably undue attention (Jones 2011), from its initial identification in Sudan the Ebola genus appears repeatedly associated with shifts in land use related to nascent capitalization (Groseth et al. 2007). Newly emergent variants appeared connected to area-specific cotton (WHO/International Study Team 1978; Baron et al. 1983), mining (Bertherat et al. 1999), and logging (Morvan et al. 2000), with each outbreak subsequently amplified in local hospitals (Mylne et al. 2014). The WHO/International Study Team (1978) began its description of the very first reported outbreak:

> In Sudan, the first cases of hemorrhagic fever are thought to have originated in Nzara township in three employees of a cotton factory situated near the town centre. The factory forms part of a large agricultural cooperative with 2000 employees. A staff of 455 is employed in the factory, which produces cotton cloth from raw cotton grown in small holdings throughout the region.

The financial capital involved seems negligible compared with what globally circulates and more recent influxes into the Sub-Sahara, but novel epizootics may arise from even small shifts in land use (Patz et al. 2004; Murray and Daszak 2013). The Zande Scheme, a colonial developmental strategy begun 1946 around forcing relocated Azande to grow cotton in Nzara's environs and spin and weave it in town, was brought to a halt by civil war in 1965 (Russell and McCall 1973/2013; Onwubuemeli 1974). But upon peace in 1972, the area rapidly repopulated and much of the local rainforest was reclaimed for subsistence farming, with cotton, as of the outbreak, continuing as the area's dominant cash crop (Roden 1974; Smith et al. 1978).

Although Smith et al. (1978) note the roof of Nzara's cotton factory housed large populations of bats, anticipating efforts to identify a source in every Ebola outbreak to follow, the epistemological implications to draw here are broader. The elusive balance between what Gonzalez et al. (2005) identify as Ebola's sudden clinical expression and silent circulation suggests the molecular, clinical, and epidemiological characteristics of the virus are necessary but insufficient explanations for disease dynamics (Engering et al. 2013; FAO 2013b; R.G. Wallace et al. 2015). The agroeconomic context within which wildlife, crops, livestock, and human populations interact appears a foundational cause for characterizing the virus's epizoonosis.

The difficulty in identifying a clear, single reservoir for Ebola in Africa (and now elsewhere) appears to represent more a paradigm's breakdown than a failure in diligence (Olival and Hayman 2014). Ebola's dynamics likely embody contingent interactions across multiple host species, communities, and environments reminiscent of avian influenza (Gilbert and Pfeiffer 2012), with repeated if relatively cryptic transmission across different host guilds in different anthropogenic settings (Plowright et al. 2015). The greater the combinatorial of host species so identified, the less explanatory power traditional modeling apparently offers, including systems of simultaneous equations tracking the transmission mechanics internal to a host population, from susceptibles to the infectious to the removed (e.g., Hayman et al. 2013). Strictly ecological niche models meanwhile omit social sources of external effect (e.g., Pigott et al. 2014). The order of complexity seems to extend beyond the approaches' methodological framework.

To more explicitly articulate the relationships among the inputs underlying pathogen emergence, including Ebola's, we expand here on R.G. Wallace et al. (2014), developing a series of stochastic models integrating epidemiology, spatial dynamics, and economics. The inductive, epiphenomenal analysis introduced here aims to guide field research in a more integrative direction across biocultural domains, even beyond what the One Health approach connecting wildlife, livestock, and human health claims sufficient (Wood et al. 2012; R.G. Wallace et al. 2015).

We begin with a simple analytic model of epidemic population growth under conditions of stochastic uncertainty. We move to modeling a pathogen's spatial dynamics under stochasticity by simple diffusion and by travel flows structured by underlying agroecologies. To introduce an agroecological logic gate for epidemic control, we characterize Ebola's emergence in another part of the global food chain–in commodity hog in the Philippines. Finally, we develop a variation of the Black–Scholes pricing model to sketch the effects of environmental stochasticity on the costs of biocontrol and containment under different agroeconomic regimens.

2.2 Modeling an Epidemic in a Stochastic Environment

Perhaps the simplest mathematical treatment of the early stages of an epidemic is found in a stochastic differential equation (SDE) (Oksendal 2010; R.G. Wallace et al. 2014).

Taking N_t as the number of infected individuals at time t, the outbreak at first will grow according to the SDE

$$dN_t = \alpha N_t + \sigma N_t dW_t \qquad (2.1)$$

where $\alpha > 0$ is the rate of growth, $\sigma > 0$ the magnitude of system noise under conditions of volatility, and dW_t represents a simple white noise process. Applying the Ito chain rule to $\log[N_t]$ to compute the derivative of a stochastic variate, the time dependence of the outbreak is driven by the expression

$$N_0 \exp[(\alpha - \sigma^2/2)t] \tag{2.2}$$

where N_0 is the initial number of infected individuals, and the term $\sigma^2/2$ represents the classic Ito correction factor.

If noise is sufficiently large, the infection population rapidly declines to zero, in spite of the condition $\alpha > 0$.

The constant $\alpha > 0$ in Eq. (2.1) represents in Kermack and McKendrick's influential deterministic, exactly solvable, and approximate susceptible-infectious-removed (SIR) model (Bailey 1975, Eq. 6.1) the basic reproductive number "R_0" factor $\beta n - \gamma$, where β is the infection rate, γ the removal rate, and n the initial uninfected population. If $\beta n - \gamma > 0$ in that formalism, the epidemic ignites and proceeds up the epicurve.

At this point, reactive control strategies aimed at decreasing β by prophylaxes, hygiene, and/or trade restrictions or at increasing the removal rate γ by vaccination or depopulation can be costly and effort-intensive in human and livestock populations, and difficult in and even detrimental to wildlife, especially for pathogens of high mutation rates and high dispersal (Capua and Marangon 2007; Kleczkowski et al. 2012; Humphries-Waa et al. 2013; Knight-Jones and Rushton 2013; Longworth et al. 2014; Zhang et al. 2014).

By contrast, the second term of Eq. (2.1), in which the noise dW_t is amplified by the factor σN_t, introduces a volatility that differentiates our approach from Kermack–McKendrick. In this treatment, environmental stochasticity is no mere fuzzy error, rather representing a keystone impact upon epidemiological systems. Sufficient noise snuffs the epidemic candle, as it were. Tornatore et al. (2005) use a similar but more complicated stochastic model, finding a similar condition for global asymptotic stability.

Indeed, such preventative control may be intentionally "farmed." Following Khasminskii (1966/2006; 2012, Theorem 4.1), the formalism in Appendix extends the model to multidimensional systems subject to nonlinear cross-influences in which relatively stable endemic modes—equilibrium states of dynamic Markov processes—can be suddenly triggered into epidemic explosions. In the other direction, we hypothesize careful internal structuring—what may be called "epidemic prevention farming"—can produce the noise sufficient enough to control outbreaks over a wide swath of epidemiological parameters.

While Kermack–McKendrick SIR models are often represented as having complex probability structure (e.g., Bailey 1975), recasting epizootics within an SDE can open up additional insight, allowing the well-understood methods of differential equations to be extended in a relatively simple manner with the Ito chain rule and related tools. The approach can often cut through considerable mathematical underbrush.

A simple spatial version of our initial model follows Okubo (1980), introducing diffusion of the form,

$$\partial N(x,t)/\partial t = \mu \partial^2 N(x,t)/\partial x^2 + \alpha N(x,t) \tag{2.3}$$

We assume "cultivation," in a large sense, takes place on spatial patches of dimension L, separated by a fixed distance across which infection must undergo spatial diffusion.

A solution using spatial Fourier series then leads to a time dependence proportional to

$$\exp[(\alpha - C^2 \mu / L^2)t] \tag{2.4}$$

where α is again growth rate, μ is the spatial diffusion coefficient, and C is a constant of order 1 that depends on the dimensionality of the diffusion.

Thus the infection dies out if L is less than the critical patch size $I_c = C\sqrt{\mu/\alpha}$. The result can be obtained directly from dimensional analysis. Indeed, such a calculation has served as the basis for the design of pest control strategies. See the classic text by Murray (1989, Sect. 14.8) for details.

How do the initial SDE and spatial diffusion model relate? Under both treatments, if ecological noise is sufficiently large, the infection population will collapse whatever its coefficient of growth. We can track the relationship deeper into the models' mechanics, with both conceptual and practical implications.

Following Tornatore et al. (2005), in Eq. (2.1) we are assuming a stronger role for volatility the σN_t term than what follows from a simple diffusion approximation to the deterministic Kermack–McKendrick approach (e.g., Tuckwell and Williams 2007). The latter generates a volatility term $\propto \sqrt{\sigma N_t}$. Indeed, such correspondences are often bounded in very particular ways. Beddington and May's (1977) map $dN_t = \alpha N_t \rightarrow \alpha(t)N_t = (\alpha_0 + \sigma dW_t)N_t$ is needed to make Eq. (2.2) consistent with the spatial analysis of Eq. (2.4), a correspondence reduction suggesting that stochastic generalizations of deterministic models are not the easy routine often assumed.

The implications extend out to how we think about the phenomena modeled. The influence of stochasticity appears to be of its own domain, distinct from factors driving deterministic population growth and explicitly dependent on underlying agroecological, epidemiological, and socioeconomic contexts, constructs, and policies, a conclusion Liu (2013) and Cai et al. (2013), modeling SIR epidemics, arrived at themselves.

R.G. Wallace et al. (2014) use the model elaborated upon here to frame the recent Ebola outbreak in West Africa, hypothesizing a key ecological role for the expansion of palm oil agroforestry from artisanal, episodic, and dispersed to year-round hybridized monocrop. Expanding patches of such commodity plantations, in conjunction with increasing deforestation, provide an attractive alternate habitat for behaviorally plastic species of fruit bats and, following Saéz et al. (2015), insectivore species thought to act as Ebola reservoirs.

The method serves as the foundation for an SDE model of noise-driven phase transitions in a physical system (Horsthemeke and Lefever 2006), suggesting a more general analysis of disease dynamics could use the noise strength σ as a "temperature" analog in modeling punctuated "socioviral" evolutionary transitions (e.g., R. Wallace and R.G. Wallace 2015).

2.3   A More Realistic Spatial Model

The spatial mechanisms used in epidemic theory, as in Eq. (2.3), are usually variants on simple diffusion. More realistic treatments examine space that is strongly structured by travel flows, which focus on biological and social processes that embody the detailed mechanisms for spatially contagious diffusion at different scales.

Gould and R. Wallace (1994) explore the spread of another infectious disease of African origins—HIV/AIDS in the USA—using empirical measures of travel intensity derived from a Markov analysis. The model is based on the area density of the equilibrium distribution of a stochastic matrix $\mathbf{P}$ indexed by exchange measures between spatial subunits that is normalized to unit row sums (Kemeny and Snell 1976). The approach leads to a canonical form related to, but significantly different from, classical diffusion theory, in that spatial homogeneity is now replaced by the complex structures of actual travel in the affected region. An important feature is the identification, by dimensional analysis, of a "characteristic area" associated with diffusion in a travel field that expands as the infection propagates.

The argument is direct. Following Gould and R. Wallace (1994), we assume a basic model

$$n_i(t) = f(t, \rho_i/A_i) \tag{2.5}$$

where i is the index of the spatial area of interest, n_i the case rate per unit population, t is the time, A_i the area of subunit i, ρ_i is the Markov equilibrium distribution associated with unit i, and f is a monotonic increasing function. The vector ρ is defined by the relation $\rho = \rho\mathbf{P}$. Since n_i is a dimensionless number, we must have an appropriate dimensionless monotonic increasing relation

$$n_i(\tau) = F\left[\frac{\rho_i}{A_i}\mathcal{A}(\tau)\right] \tag{2.6}$$

where $\tau \equiv t/T_0$, T_0 is a characteristic system time, and $\mathcal{A}(\tau)$ is a characteristic area that grows as the disease spreads. The generalization to a continuous system would be direct, involving introduction of spatial variates $\mathcal{X}$, and functions

$$n(\tau, \mathcal{X}), d\rho/dA|_{\mathcal{X}}$$

F may actually be fairly complex, incorporating dimensionless "structural" variates for each subdivision i, for example, local percentage indices of income stratification or occupational status, crime rates, etc., in the case of R. Wallace et al. (1997).

We conjecture, however, that $A(\tau)$ itself expands according to a stochastic diffusion process, even though the process defined by F is a kind of deterministic mixmaster driven by systematic local travel patterns. From the perspective of the polio examples in R. Wallace and R.G. Wallace (2015), it may be necessary to treat the spread of the characteristic area as a carrying capacity problem, since, once the infection has become endemic, it is "everywhere." The appropriate SDE follows as

$$dA_\tau = [\alpha A_\tau (1 - A_\tau/K)]d\tau + \sigma A_\tau dW_\tau \qquad (2.7)$$

which generalizes Eq. (2.1) to a carrying capacity K, in the presence of white noise.

Using the Ito chain rule on $\log(A)$, one obtains—as a consequence of the added Ito correction factor—the long-time endemic limits

$$A \to 0, \quad \alpha < \frac{\sigma^2}{2}$$

$$A \to K\left(1 - \frac{\sigma^2}{2\alpha}\right), \quad \alpha \geq \frac{\sigma^2}{2} \qquad (2.8)$$

Thus, in this model, the "noise" σ can depress the final endemic level of infection even for a propagating epidemic.

Figure 2.2 shows two simulation examples, with σ below and above criticality. For the first, the characteristic area fluctuates about a lowered endemic level. For the second, the area compresses toward extirpation.

R. Wallace et al. (1999) analyze data for very large-scale AIDS spread in the USA, and other contagious processes, using related Markov methods. The team found HIV diffused down the hierarchy of US cities—from large economic centers to second cities—by way of the structure of national travel as well as the policies of deindustrialization and inner city discrimination driving unsafe behaviors spreading HIV. The approach may be adaptable to slower systems that do not equilibrate.

Transitions in Ebola's characteristic area may be defined by a similarly broad combination of factors, in this case including migration patterns and agroeconomic shifts in land use across wildlife and human populations. Developing industrial agriculture influences not only the epizootic interface across species, but also road networks, cheap transport, and the ecoimmunology of human populations subjected to shifts in labor requirements, demographics, diets, and associated habits of behavior.

Should ZEBOV be characterized in this way by noise σ below criticality, the recent contraction in the outbreak in West Africa may be short-lived, perhaps even in the face of welcome changes in public health intervention (Washington and Meltzer

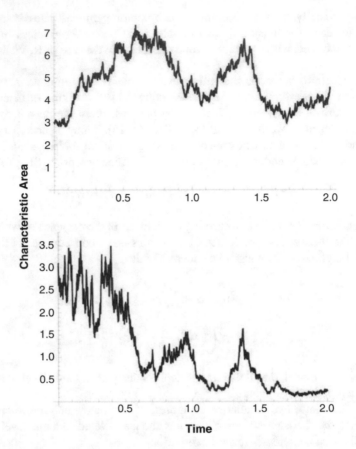

Fig. 2.2 Simulating $\mathcal{A}(\tau)$ based on the Ito chain rule expansion of $\log(\mathcal{A}_\tau)$ using Eq. (2.7). The simulations apply the ItoProcess function in Mathematica 10 for white noise. $\mathcal{A}(0) = 3$, $K = 10, \alpha = 1, \sigma = 0.5, 1.5$. The critical value for σ is $\sqrt{2}$. 2000 time steps. While the *upper trace* fluctuates at values less than K, the *lower* collapses toward zero.

2015; Merler et al. 2015; WHO 2015). Under such circumstances, ZEBOV may now be endemic in human and reservoir populations alike, prone to repeated spillover and rebound.

There appears something fundamental still missing in the catalog of mechanisms by which a marginal epizootic embedded in deep agroforestry turns into a regionalized urban epidemic. To start the effort at more explicitly characterizing the interactions influencing the virus's shift across biocultural domains, we next review another Ebola species' emergence. Anthropogenic impacts on its "wild" reservoir's habitat appear to have brought the comparatively fast-evolving *Reston ebolavirus* into its own human-proximate host, ecology, and molecular phenotype (Carroll et al. 2013).

2.4   REBOV in Commodity Hog

Ebola has entered the global food chain by means other than neoliberalizing horticulture in Sub-Saharan Africa.

In 2008, half-a-world away in the Philippines, in an area that Peterson et al. (2004) projected as a suitable Ebola niche, *Reston ebolavirus* emerged twice in hog raised in industrial filieres, outside Metro Manila (1.7 million people) and Dagupan City (150,000) (Barrette et al. 2009; Miranda and Miranda 2011; Sayama et al. 2012) (Fig. 2.3a). The REBOV outbreaks followed several in cynomolgus macaques in the Philippines between 1989 and 1996, although the virus's regional origins remain obscured as purifying selection may mask the phylogenetic age of ebolavirus lineages and filovirus-like elements have been detected in bat and other small mammal genomes dating back tens of millions of years (Taylor et al. 2010; Wertheim and Pond 2011).

The Win Farm in Sto. Nino, Pandi, Bulacan, one of the piggeries on which REBOV emerged in 2008, is 30 km from the Diliman Arboretum Forest and 125 km from the Quezon National Forest Park (Fig. 2.3a). One apparent natural reservoir, the wide-ranging karst fruit bat *Rousettus amplexicaudatus*, was found IgG ELISA-positive for REBOV NP in the two parks (Heaney et al. 2005; Cardinoza and Reyes 2009; Taniguchi et al. 2011). The highly vagile *Rousettus* forages widely across orchards, other agricultural areas, and disturbed secondary forest.

The outbreaks among lab *Macaca* in Reston, Texas, Sienna, and the Philippines were traced to a single monkey farm and exporter in Calamba, Laguna (Miranda and Miranda 2011) (Fig. 2.3a). The 2008 outbreak showed REBOV more widely, if heterogeneously, distributed (Sayama et al. 2012). At the Sto. Nino farm, 79 % and 81 % of the swine tested positive for REBOV-NP and -GP specific antibodies by IgG-ELISA. At the second piggery hit, the Lambino farm in periurbanized Barangay Parian, Manaoag, Pangasinan, 90 % and 67 % tested positive. None tested positive on a Tarlac farm sampled 2 years later, following negative tests at two inspection check points in St. Nino, San Jose City, Nueva Ecija (Barrette et al. 2009; Sayama et al. 2012).

The 2008 swine strains shared 95 % similarity and the same clade with those of the macaque outbreaks, but significantly greater genetic divergence from each other than from the 1989 reference isolate, suggesting both continual circulation and polyphyletic origins (Barrette et al. 2009). Multiple substrains appear endemic and the 80 % prevalence suggests either broad environmental exposure or hog-to-hog transmission. Barrette et al. (2009) also identified six clinically unaffected farm workers (4 %) who, denying all contact with bats and monkeys, expressed IgG antibodies to REBOV, confirming human infection (Bausch 2011).

Hog comprises 60 % of total Filipino meat output, at 24.5 million head and 956,250 tons annual live weight (Costales et al. 2007; Lapus 2014). Forty-seven percent of production takes place on the island of Luzon where the 2008 outbreaks emerged in Bulacan and Pangasinan, the provinces hosting the country's two greatest concentrations of commercial hog (Costales et al. 2003; Alawneh

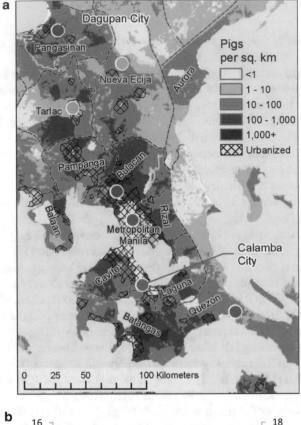

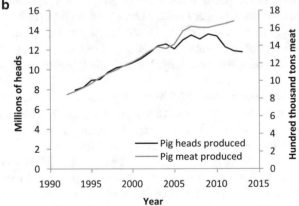

Fig. 2.3 REBOV in the Philippines. (**a**) Hog farm outbreaks, Pandi, Bulacan and Manaoag, Pangasinan, 2008 (*purple*); negative hog tests, Tarlac City and St. Nino, San Jose City, Nueva Ecija, 2008–2010 (*green*); positive IgG ELISA in Geoffroy's rousette bat *Rousettus aegyptiacus*, 2008–2009 (*blue*); positive cynomolgus macaques *Macaca fascicularis*, 1989–1996 (*pink*); on Livestock Geo-Wiki distribution of pigs in Central Luzon, Philippines, 2006 (pigs/km^2) (Barrette et al. 2009; Taniguchi et al. 2011; Miranda and Miranda 2011, Sayama et al. 2012; Robinson et al. 2014). *Projection*: Cylindrical equal area with a standard parallel of 10 degrees. (**b**) Hog production in the Philippines. Pig heads and pig meat produced, 1992–2013. UNFAO data.

et al. 2014). By the Import Liberalization Program, Ginintuang Masaganang Ani Livestock Program, the Agriculture and Fisheries Modernization Act, and World Trade Organization obligations for minimum access volume for multinational hog producers, a rural, smallholder-based sector on the island has been transformed into conurbanized chains of advanced commercial operations, doubling production (David 1997; Kelly 2000; Delgado et al. 2003; Costales et al. 2003; Bello 2003; Verburg and Veldkamp 2004; Costales et al. 2007; Stanton et al. 2010) (Fig. 2.3b). Hog inventories are growing at 4.5 % a year on Luzon, and, as in the USA and other industrial countries, farm numbers there are declining and heads-per-farm increasing (Catelo et al. 2008). Production continues to consolidate and capitalize with an eye toward exporting output across the ASEAN Free Trade Area.

Kelly (1998) describes the resulting shifts in landscape,

> Large swathes of irrigated agricultural land in the 'rice bowl' provinces of the Central Luzon and Southern Tagalog regions have been converted to a variety of urban and industrial uses: export processing zones and industrial estates; institutions such as hospitals and universities; leisure landscapes such as golf courses, resorts and theme parks; and, most significantly in terms of the area involved, residential sub-divisions. The result is a reworking of the social and economic, as well as the physical, landscape of formerly 'rural' areas, such that even within the same household the urban-industrial economy might co-exist with agricultural production.

Land conversion, commoditized food chains, and deforestation have broadened the interfaces wild REBOV reservoirs, livestock, and human populations apparently share.

Shively and collaborators (Coxhead et al. 1999; Shively 2001; Coxhead and Jayasuriya 2002; Shively and Pagiola 2001, 2004) have argued policies of import substitution industrialization, aimed at replacing foreign imports with domestic production in capital-intensive sectors, have by their weak growth and underemployment prompted increasing smallholder land colonization, agricultural intensification, and deforestation. But expanding imports, land leasing, and foreign direct investment; lowering duties on corn for internationally connected hog producers; and contract husbandry have together proven no panacea either. Farm consolidation, differential farmgate pricing structured by the WTO Agreement on Agriculture at smallholder expense, and 'flexible' zoning for politically connected land expropriation have had detrimental impact on the environment, especially in Central Luzon (Kelly 1998, 2000; Pagiola and Holden 2001; Costales et al. 2003; Bello 2003; Habito 2011; Kelly 2011; Borras and Franco 2011). The new economic paradigm has spurred speculation on both marginal arable land and primary forest; expanded agriculture into new zones of production; and driven struggling smallholders into cities, pluriactive cycle migration, or further into what forest remains.

Behaviorally plastic frugivore bats are documented to roost in plantation farms that have deforested or disturbed traditional habitats, a transition linked to pathogen spillover into human and hog alike (Chua 2003; Halpin and Mungall 2007; Luby et al. 2009, Leroy et al. 2009, Shafie et al. 2011). Deforestation is particularly severe in the Philippines, producing limited vegetative cover and patchy food distributions for fruit bats, mitigated in part by among other crops oil palm, ethanol cassava, taro, coconut, and orchards grown directly on livestock operations for fodder and shade (Moog 1991; Shively 2001; Verburg and Veldkamp 2004; Heaney et al. 2005, Jones et al. 2009, Borras and Franco 2011; Sedlock et al. 2008).

By this point REBOV may also be endemic to the regional hog food chain, alongside a growing list of pathogens, including *Actinobacillus pleuropneumoniae*, atrophic rhinitis, classical swine fever, *Haemophilus parasuis*, *Mycoplasma hyopneumoniae*, porcine circovirus type 2, type 2 porcine reproductive and respiratory syndrome virus, pseudorabies virus, swine influenza virus, hepatitis E, and a variety of septicemia, hemophillosis, coccidiosis, nematodiasis, gastroenteritis, and parvoviral infections (Tateyama et al. 2000; Drew 2011; Khan et al. 2013; Alawneh et al. 2014; Ng and Rivera 2014).

REBOV was recently reported in commercial hog in China, where half of the world's domesticated pigs live. Pan et al. (2014) report REBOV detected on three hog farms 35 km from each other in Shanghai, sharing 96–99 % sequence similarity with two Philippine variants and the Manaoag strain the Shanghai isolates' immediate basal ancestor. The recycled inputs, overlapping cohorts, and extensive value-added networks of typical regional hog production may promote infection across farms, even within the all-in/all-out model of production (Graham et al. 2008; Atkins et al. 2010; Otte and Grace 2013). Bausch (2011) additionally hypothesized refrigeration associated with commercial hog processing could preserve the fragile virus. On the other hand, REBOV has also been reported in China in Leschenault's rousette fruit bats (*Rousettus leschenaulti*) (Yuan et al. 2012).

The two epizootic transitions—repeated spillover across a wider agroecological interface and endemicity in the hog chain—need not be mutually exclusive. Carroll et al. (2013) hypothesize REBOV underwent a genetic bottleneck in the Philippines consistent with both a decline in a host reservoir driven by deforestation and a founder event in newly commercialized hogs.

2.5 Pandemic Ebola by Way of Commodity Livestock?

What of any direct epidemiological fallout, other than the potential economic impacts of such a new agroepizoology? REBOV has infected humans handling hogs along the value-added chain, but to date to no clinical danger or measurable human-to-human transmission.

On the other hand, the ZEBOV outbreak in West Africa signaled virulent variants can shift from limited spillovers in deep forest to regional, and potentially pandemic, H2H spread. Indeed, human history is marked by such epidemiological transitions. Diphtheria, influenza, measles, plague, pertussis, tuberculosis, cholera, HIV/AIDS, dengue, malaria, and yellow fever, among other examples, are all human diseases that originated in shifts in host reservoir, characteristic area, functional ecology, and/or modes of transmission (Pearce-Duvet 2006; Wolfe et al. 2007; Kock et al. 2012; R.G. Wallace et al. 2015).

In that context, a growing body of experimental research is closely exploring the extent to which Ebola can transmit from hogs to humans. Marsh et al. (2011) challenged 5-week-old pigs with 10^6 TCID50 of REBOV-08 by the oronasal route and, in a separate experiment, subcutaneously. The team detected viral shedding from the nasopharynx 2–8 days post-exposure in pigs with subclinical infection and 6–8 days from the rectum, with infiltration across most other organs and tissues, producing gross abnormalities in lymphoid and respiratory systems.

Kobinger and Weingartl's group (Kobinger et al. 2011; Weingartl et al. 2012; Nfon et al. 2013; Weingartl et al. 2013) showed ZEBOV can infect and sicken pigs, which can transmit the virus to cynomolgus macaques—standing in for humans—without direct contact, by airborne or sapronotic transmission. The pigs suffered a respiratory syndrome that can be mistaken for other porcine diseases, characterized by the dysregulated recruitment of non-infected lymphocytes, upregulated cytokines, proapoptotic induction, pulmonary consolidation, and hemorrhagic alveolitis. In contrast, the macaques presented no respiratory symptoms but suffered systemic collapse associated with hemorrhagic fever and a dysregulated cytokine response.

Even as ZEBOV is clearly the more virulent of the two Ebola types, REBOV infections are nearly as deadly when hogs are coinfected with other pathogens associated with the commodity chain, including off the long list in the previous section, PRRSV and porcine circovirus type 2 (Barrette et al. 2009; Carroll et al. 2013). Other combinations of porcine respiratory co-infection have been shown to induce synergistic morbidity (Stark 2000; Neumann et al. 2005). The molecular mechanisms remain to be worked out, but rich networks of molecular reciprocal activation have been identified in other pathogen guilds, including between HIV and many of its opportunistic infections (Lehrnbecher et al. 2000; Lawn 2004; Guo et al. 2004; Sun et al. 2005).

The steps in molecular evolution on one possible path to REBOV pathogenicity were recently mapped. Pappalardo et al. (2016) computationally analyzed 196 Ebolavirus genomes for Specificity Determining Positions across the virus's nine proteins, including protein–protein interface sites and enzyme active sites, that differentiate the Ebola species. REBOV, the only species non-pathogenic to humans to date, differed from its pathogenic counterparts in a GP SDP and multiple SDP across structural proteins VP24, VP30, VP35, and VP40.

Pappalardo et al. noted eight structurally identifiable SDP differed across Ebolavirus in membrane-associated protein VP24, which is involved in the formation of the viral nucleocapsid and regulating viral replication. The protein is an antagonist to host interferon, binding to karyopherins $\alpha1, \alpha5$, and $\alpha6$, and critical in Ebola adaptation to novel hosts. The team found seven SDPs on the same VP24 face, implying an interface with other viral proteins or host cell. Mutations to two SDPs have been shown to block KPNA5 ($\alpha5$) interactions, critical to host adaptation, and are adjacent to a mutation that reduces interferon antagonism (Xu et al. 2014; Ilinykh et al. 2015). A third replacement a hundred amino acids down breaks a hydrogen bond, further destabilizing the protein. REBOV pathogenicity appeared impaired on all accounts here.

Pappalardo et al. account for other species-defining differences. In transcriptional co-factor VP30 the removal of two hydrogen bonds in C-terminal dimer formation and the burial of two functional residues may change the balance of transcription/replication in REBOV. Among its multiple functions, VP35 antagonizes interferon signaling by binding to host dsRNA. REBOV VP35 is overall more stable, with a reduced affinity for dsRNA. Pappalardo et al. show two SDP here: one near the linker, the other at the dimer interface, the latter shortening VP35's aspartate side chain, in all likelihood increasing the distances among nearby residues sharing hydrogen bonds, destabilizing the dimer complex. One SDP introduces backbone flexibility and destabilization to the octameric form of VP40 involved in viral transcription. Another SDP likely breaks and shortens one of VP40's helices and changes its hydrophobic core.

The authors conclude REBOV is only a few SDP shifts from evolving human pathogenicity. Although virulence may arise by any number of means, the comparison maps one short 'path to victory' for a dangerous REBOV.

The results so far together suggest virulent Ebola and human spillover across a global hog value chain expanding in size and extent a distinct possibility (Drew 2011) (Fig. 2.4a). As Marsh et al. (2011) conclude of REBOV:

> The evidence of virus shedding and replication of virus in internal organs [of swine] in the absence of clinical disease represents a potential source of infection to farm, veterinary, and abattoir workers. This appears to be an unprecedented emergence of filovirus infection in a new host that may have important biosecurity implications for both livestock health and emergence in the human food chain. Although REBOV has not been seen to result in any human disease, the basis for this observed attenuation remains unknown. The consequence of REBOV becoming pathogenic in humans is serious, and ongoing undetected infections and replication in pigs and other animals with REBOV may result in the emergence of viruses that are more pathogenic in humans and/or livestock.

Bausch (2011) adds:

> ...EBOV pathogenicity can be enhanced by serial passage in animals and cell culture...A similar result through unintentional serial passage in pigs is not out of the question...
> [T]he laboratory findings of Kobinger et al., combined with the previous results of field investigations in the Philippines...highlight the possibility of EBOV as a foodborne pathogen. This is cause for consideration, further scientific study, and prudent surveillance and prevention measures in the livestock industry in implicated areas of the world.

These preliminary results suggest the epidemiological stakes are high indeed. Multiple modes by which animal and public health might successfully intervene should be explored. Further study, then, especially in the light of environmental stochasticity's intrinsic impact on epidemic growth, need be extended to the broader agroeconomic context out of which new Ebola variants emerge and, as we explore in the next section, may perhaps be controlled.

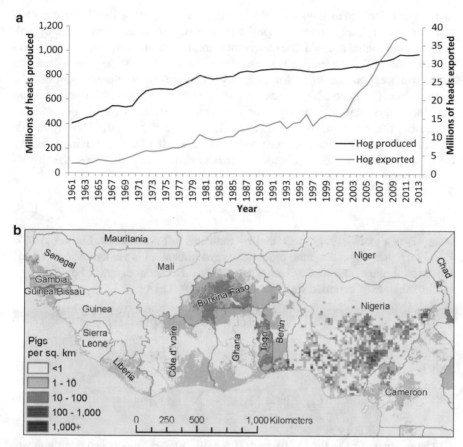

Fig. 2.4 Expanding hog sector. (**a**) Global hog production and international exports, 1961–2013. UNFAO data. (**b**) Distribution of hog across West Africa, 2006 (pigs/km-sq) (Robinson et al. 2014). *Projection*: Cylindrical equal area with a standard parallel of 10°. East Africa–across Uganda, Rwanda, and parts of Kenya–is characterized by comparable intensification.

2.6   An Agroecological Logic Gate for Epidemic Control

In an effort to reimagine the agroeconomic geographies that propagate pathogens, as Ebola apparently instantiates in West Africa and the Philippines, we need aim to empirically define a critical ecosystemic "temperature" at which an outbreak is instead "sterilized". Specifically we need to incorporate measures of the emergent "friction" that agroeconomic and policy barriers impose upon propagating pathogens (Levin et al. 1998; R. Wallace 2002; Meentemeyer et al. 2012; Boyd et al. 2013).

There are complications, however. As discussed earlier, the noise terms in Eqs. (2.1) and (2.7), the dW_t, are not necessarily simple white noise characterized by a single intensity parameter. Indeed, the relatively straightforward treatment of fractional Brownian motion in R.G. Wallace et al. (2014) implies that such a "temperature" is likely to be a composite function of the intensity and color of both temporal and spatial noise. Some conceptual simplification, however, is possible via an extension of the embedding mathematics to include Levy-type jumps—random walks with step-lengths drawn from a heavy-tailed probability distribution.

Equations (2.1) and (2.7) are expressed in terms of classic white noise having the simple quadratic variation $[W_t, W_t]_t = \sigma^2 t$, where σ is the single available parameter, the noise magnitude (Oksendal 2010; Protter 1990). The arguments of Eqs. (2.1) and (2.7) can be extended to other kinds of noise, having arbitrary quadratic variation and discontinuous stochastic jumps, since they have the characteristic form

$$dZ_t = Z_{t-} dY_t \tag{2.9}$$

where Y_t is a stochastic process, and $t-$ indicates left-continuous. Letting $\Delta Y_t = Y_t - Y_{t-}$, representing the jump process, the generalization is via the Doleans–Dade exponential, the solution of a class of stochastic differential equations defined by a semimartingale of bounded variation (Protter 1990). Then

$$Z_t = \exp(Y_t - 1/2[Y_t, Y_t]_t^C) \Pi_{s \le t}(1 + \Delta Y_s) \exp(-\Delta Y_s) \tag{2.10}$$

where $[Y_t, Y_t]_t^C$ is the path-by-path continuous part of the quadratic variation of Y_t. This is written as

$$[Y_t, Y_t]_t^C = [Y_t, Y_t] - \sum_{0 \le s \le t} (\Delta Y_s)^2 \tag{2.11}$$

The product term in Eq. (2.10), with (Levy-like) jump processes having nonzero Δ, converges.

The essential point emerging from the formalism is that, if the structurally imposed environmental stochasticity $[Y_t, Y_t]_t^C$ is monotonically increasing in time at a greater rate than Y_t, the exponential factor $\exp(Y_t - 1/2[Y_t, Y_t]_t^C)$ in Eq. (2.10) drives the system toward extinction. The Appendix examines more complicated multidimensional processes in which (relatively) low level endemic infection can be either driven to epidemic explosion or, the better outcome, stabilized.

Setting Levy-like jumps aside for the moment, it is possible to extend Eq. (2.9) as

$$dZ_t = dL_t + Z_t dY_t \tag{2.12}$$

where L_t is a generic stochastic "seeding" process, here representing a pathogen's incursion from outside the system of interest. Following Protter (1990), if the stochastic "frictional" variability represented by the quadratic variation $[Y_t, Y_t]$ is sufficiently large, there will be no explosion of infection, which will, at worst, asymptotically converge on the imported cases, i.e., $Z_t \to L_t$.

This is a striking outcome in a broadly comprehensive model.

Remarkably, these results can be interpreted in terms of the Data Rate Theorem, which imposes necessary conditions on information transmission and links control theory with information theory:

An outbreak can be conceptualized as a signal sent through a channel, whose topology is embodied by the local ecology. Environmental stochasticity—noise against the signal—can disrupt the broadcast. That is, $d[Y_t, Y_t]^C/dt$ (or, for Eq. (2.12), $d[Y_t, Y_t]/dt$) can be viewed as an imposed control signal on a highly nonlinear "pathogen factory" analog logic gate.

A logical operation, controlled here explicitly by agroeconomic policy, is performed on multiple inputs, in this case a network of agroecological relationships, to produce a single logical output, here the state of an epizootic population.

The Data Rate Theorem, a generalization of the classic Bode Integral Theorem for linear control systems, describes the stability of feedback control under data rate constraints (Nair et al. 2007). What is the smallest feedback data rate above which such an unstable dynamical system can be stabilized? In our context, what is the smallest amount of policy-determined environmental stochasticity we need to stabilize or prevent an expanding outbreak?

Given a noise-free data link between a discrete linear "factory" and its controller, unstable modes can be stabilized only if the feedback data rate $\mathcal{H}$ is greater than the rate of "topological information" generated by the unstable system. For the simplest incarnation, if the linear matrix equation of the "factory" is of the form $x_{t+1} = Ax_t + \ldots$, where x_t is the n-dimensional state vector at time t, then the necessary condition for stabilizability is

$$\mathcal{H} > \log[|\det A^u|]   \tag{2.13}$$

where det is the determinant and A^u is the decoupled unstable component of A; that is, the part with eigenvalues ≥ 1. Thus there is a critical positive data rate below which there does not exist *any* quantization and control scheme able to stabilize an unstable system. Here, $\mathcal{H} \equiv 1/2d[Y_t, Y_t]^C/dt$ (or $1/2d[Y_t, Y_t]/dt$), and a sufficient "data rate" condition can be expressed, applying the mean value theorem for monotonicity, as

$$\mathcal{H} = \frac{1}{2}d[Y_t, Y_t]^C/dt > dY_t/dt   \tag{2.14}$$

A somewhat different approach, deriving a policy driven, fully information-theoretic $\mathcal{H}$ via large deviations theory, is outlined in Appendix.

As REBOV's origins in the Philippines clearly demonstrate, the dynamics of infection initiation and spread—including structured spatial diffusion and stochastic jumps—may be quite complicated. But a sufficient agroecological control signal, produced by adequate variation in space, time, and mode of production—high enough "noise"—will almost always stabilize and control pathogen outbreaks, largely limiting them to either imported cases or intermittent spillovers that burn out in the face of the area's ecological resilience and/or public health capacity

(Hogerwerf et al. 2010; Lewnard et al. 2014). However frequent or deadly such outbreaks—think up to this point Ebola in Central Africa—they are largely constrained to local domains.

2.7   The Costs of Biocontrol and Containment

As our description of REBOV in the Philippines intimated, the crux of epidemic control clearly extends out beyond susceptible-infectious-removed dynamics, the effects of the microeconomics of livestock production on disease (e.g., Boni et al. 2013; Allen and Lavau 2014), or even the agroeconomically informed spatial modeling with which we began, however, illuminating all such modeling may be. Causality extends out into the way the fabric of the global economy interpenetrates regional agroepizoology (Wilson et al. 1994; R. Wallace and R.G. Wallace 2015; R.G. Wallace et al. 2015).

We take a step in that direction, relating epizoology and economic policy in a more explicit fashion. We apply the Black–Scholes approach to option pricing in finance toward modeling the cost in resources needed to control the agroecological logic gate introduced in the previous section (by which pathogen populations explode or stabilize) (Black and Scholes 1973).

Let $\mathcal{H}_t$ be the value of $\frac{1}{2}d[Y_t, Y_t]^C/dt$, our environmental stochasticity, or the value of the rate function in a large deviations analysis of the Mathematical Appendix at time t. Under conditions of both white noise and macroscopic volatility, a general relation can be written as

$$d\mathcal{H}_t = f(t, \mathcal{H}_t)dt + b\mathcal{H}_t dW_t \tag{2.15}$$

where the magnitude of the noise *in this iterated system* is now expressed as b to avoid confusion with the earlier development.

Let $M(\mathcal{H}_t, t)$ represent the rate of available resources—direct short- and long-term financial investment plus "opportunity costs" from those expenditures—needed to achieve $\mathcal{H}_t$ at time t, and expand using the Ito chain rule,

$$dM_t = \left[\partial M/\partial t + f(\mathcal{H}_t, t)\partial M/\partial R + \frac{1}{2}b^2\mathcal{H}_t^2\partial^2 M/\partial\mathcal{H}^2 \right] dt$$
$$+ [bR_t\partial M/\partial\mathcal{H}]dW_t \tag{2.16}$$

As in the original Black–Scholes model—which uses different terminology—we define $\mathcal{L}$ as the Legendre transform (Pettini 2007) of M, taking the involutive transformation of function M to produce

$$\mathcal{L} = -M + \mathcal{H}\partial M/\partial\mathcal{H} \tag{2.17}$$

Using the heuristic of replacing dX with ΔX in these expressions, and applying the results of Eq. (2.16), gives the relation

$$\Delta \mathcal{L} = \left(-\partial M/\partial t - \frac{1}{2}b^2 \mathcal{H}^2 \partial^2 M/\partial \mathcal{H}^2 \right) \Delta t \qquad (2.18)$$

Analogous to the classic Black–Scholes calculation, the terms in f and dW_t cancel out, so that the effects of noise are subsumed in the Ito correction involving b. This invokes powerful assumptions of regularity that may be violated. Matters then revolve about model robustness in the face of such violation.

As the Legendre transform of M, $\mathcal{L}$ is a kind of entropy that can be expected to reach a constant rate of production at nonequilibrium steady state (nss). Then $\Delta \mathcal{L}/\Delta t = C \geq 0$, $\partial M/\partial t = 0$, so that

$$-\frac{1}{2}b^2 \mathcal{H}^2 \partial^2 M/\partial \mathcal{H}^2 = C \qquad (2.19)$$

The solution is

$$M_{\text{nss}} = \frac{2C}{b^2} \log[\mathcal{H}_{\text{nss}}] + \kappa_1 \mathcal{H}_{\text{nss}} + \kappa_2 \qquad (2.20)$$

at nonequilibrium steady state.

In this form the treatment suggests two basic policy options. If $\kappa_1 = 0$, then *the cost of epidemic control grows only as the log of the policy-driven variate* $\mathcal{H}$. If $\kappa_1 > 0$, then the cost will be dominated by linear growth in $\mathcal{H}$.

The inference is that the overall financial cost of epidemic prevention and control depends on agroeconomic policy's impact on environmental stochasticity. In the context of this analysis, regional planning that introduces "frictional" variation in space, time, and mode on agroecology, raising b in Eq. (2.20), controls costs specific to an outbreak at a rate much lower than linear. In contrast, the biocontainment option, represented by the linear-dominated form of Eq. (2.20), produces a very large constant of proportionality, κ_1.

The inherently explosive epizoologies of commodity agricultures—however, frequently biocontained—appear exorbitantly expensive as a matter of first principle.

2.8   Discussion and Conclusions

The ZEBOV virus behind West Africa's Ebola outbreak appears a commonplace phenotype, with a typical case fatality rate, incubation period, and serial interval (WHO Ebola Response Team 2014). How, then, to explain the unprecedented outbreak? While the virus did not fundamentally change, Africa did (Gatherer 2015). The region's economically driven transformations in land use appears to have changed the agroeconomic matrices through which environmental stochasticity acts as an inherent brake upon pathogen momentum at the population level (Bausch and Schwarz 2014; R.G. Wallace et al. 2014). Commodity agriculture's spatial and functional expansions may by virtue of deforestation and monocultivation destroy many a pathogen. But in stripping out the agroecological "friction" diverse functional geographies impose on systems of potential susceptibles, such production may liberate many another pathogen, especially those circulating among reservoir hosts that adapt to the new agriculture (e.g., monkeys, birds, and bats).

The outbreak of REBOV in hog in 2008, marking another such agroecological transition, was initially greeted with as vociferous if less widespread public and scientific alarm (WHO 2009; Cyranoski 2009). While public interest subsequently receded, researchers across disciplines began a number of research lines carefully investigating REBOV's molecular, clinical, and epidemiological courses (Barrette et al. 2009; Bausch 2011; Kobinger et al. 2011; Marsh et al. 2011; Miranda and Miranda 2011; Sayama et al. 2012; Weingartl et al. 2012; Nfon et al. 2013; Pappalardo et al. 2016). The reports are punctuated with urgent if also characteristically conditional warnings as to both the virus's endemicity in the food chain and its potential for evolving virulent human-specific phenotypes.

By a series of stochastic models we aimed here at integrating the problematics REBOV and other Ebola species present across biocultural domains, including their molecular, geographic, and economic contexts. The punctuated emergence of a spreading pathogen via the operation of the kind of policy-driven biological logic gate we inferred here may affect not only the timing and extent of individual outbreaks, as in R.G. Wallace et al. (2014), but may embody a generalized pathogen response to agroecological pressures that, in addition to shifts in spatial patterns, select for new ecotypes and modes of transmission.

The impacts appear to extend to the financial costs of intervention. Systems that minimize environmental stochasticity's impact upon pathogen population growth incur explosive costs when new variants successfully emerge. Whether such a framework can explain any single outbreak remains to be tested. For instance, Bartsch et al. (2015) estimate that the direct societal costs of all Ebola cases in Guinea, Liberia, and Sierra Leone through mid-December 2014 range from USD $82 million to $356 million. That is a large sum, considering West Africa's exchange rates. In effect, at the risk of reductionist utilitarianism, area-specific environmental stochasticity may represent another valuable ecosystem service humanity is presently scuttling in favor of short-term earnings, one forest plot at a time (Farber et al. 2002; Zhang et al. 2007).

The model series we present here is inductive in nature—an important caveat. In the spirit of mathematical ecologist Evelyn Pielou (1977), the models aim at helping raise questions that appear largely absent from the disease literature. Specifically we aim at identifying the necessary conditions under which broad classes of zoonotic pathogens respond to anthropogenic shifts. The simplicity of the models presented here speaks to general conditions across systems, regardless of their biologies. Namely, pathogen fitness is *fundamentally* integrated with its population biological (and sociological) context. Agroeconomy is an epidemiological cause of foundational impact and no mere second-order complication (Hinchliffe 2015).

Ostensibly adjunct models could be conditionalized in various ways for testing specific systems, including, in this case, REBOV's statistical economic geographies and social phylogeographies, analyses of the value chains the virus passes through, dynamic modeling of its socio-ecological niches, an economically informed life history model of the evolution of its virulence, cost differentials across interventions, and the Markov travel analysis that we sketched around Eq. (2.5) (e.g., R. Wallace et al. 1999; Mayer 2000; R.G. Wallace 2004; Walsh et al. 2005; Atherstone et al. 2014; Lewnard et al. 2014; Pigott et al. 2014).

Following R.G. Wallace et al. (2014) and R. Wallace and R.G. Wallace (2015), one might use a number of measures of σ in Eqs. (2.2) and (2.7), of the structure of environmental stochasticity $[Y_t, Y_t]_t^C$ in Eq. (2.10), or $[Y_t, Y_t]$ associated with Eq. (2.12). These could include neoliberal or neocolonial expropriation, or the elimination of traditional farming strategies that previously isolated pathogens from livestock, wildlife, and humans at regional spatial scales. See R. Wallace (2014) for a more general exploration of canonical failure modes afflicting what are essentially cognitively "farmed" systems, here with a "crop" represented by an epidemic outbreak that one wishes to minimize. R. Wallace (2015) reexpresses these considerations explicitly in terms of an analog biological logic gate or Boolean-like circuit.

One might operationalize such broadly painted political economies in absolute geographies such as hectares leased to multinationals per total agricultural area, the number of farms consolidated in the past 10 years, and agricultural exports. On the other hand, agroeconomics are marked by relational geographies across industrial sectors (Bergmann 2013a,b; R.G. Wallace et al. 2015). Landscapes are entrained by transnational commodity chains and circuits of capital, including financial and productive circuits, with critical local effects. Products from globalized croplands, forests, or pastures eventually contribute to consumption or capital accumulation in other countries. Other landscapes are enmeshed primarily within local circuits of production and exchange.

Bergmann (2013b) extends analysis beyond characterizing landscapes that directly produce traditional agricultural exports to identifying the forests and fields that are part of commodity webs supporting export-oriented development, producing goods or services for international markets. He further differentiates foreign consumption/accumulation of "direct" agricultural goods; processed agricultural goods; manufactured goods as far afield as electronics and vehicles; and services, including air transport, insurance, and education. How such circuits of capital structure disease ecologies is a matter of ongoing research.

Such broad impact might produce large-scale ecosystemic shifts that set off spikes in pathogen population or in their cladogenesis, inducing new patterns of transmission, virulence, and endemicity (e.g., Carroll et al. 2013). That is, changes in policy or socioeconomic structure can trigger a large-scale biological logic gate and "desterilize" a natural or human ecosystem in which a pathogen has been traditionally held at a low endemic level, or simply had not previously evolved.

REBOV's emergence may represent such a canonical example. Luzon's agroeconomic geography appears to have undertaken a phase change from rural, smallholder-based production into a conurbanized daisy chain of advanced commercial operations that doubled island hog (Kelly 2000; Delgado et al. 2003; Costales et al. 2003, 2007; Catelo et al. 2008). In conjunction with deforestation, such a vertically integrated expansion appears to have broadened the interface livestock and REBOV-bearing bats share (Shively 2001; Sedlock et al. 2008; Verburg and Veldkamp 2004; Heaney et al. 2005; Borras and Franco 2011; Sedlock et al. 2011; Carroll et al. 2013). As per Bergmann et al.'s program, such a modern-day disease ecology is likely neither unconnected from the rest of the world nor erased by globalization (R.G. Wallace et al. 2010). The socioecological environments out of which REBOV and other Ebola species are evolving into food commodity pathogens are often the complex and layered products of past and present and of global and local impacts.

Indeed, a growing array of pathogens appear characterized by such multi-order anthropogenic origins, often deeply embedded within neoliberal capitalism (R.G. Wallace et al. 2015). Chua (2003) describes a similar if locale-specific transition for Nipah virus in Malaysia:

> ...[A]vailable data...suggest that a complex interplay of multiple factors led to the spillage of the virus from its natural reservoir host into the domestic pig population with subsequent spread to humans...[O]ver the last two decades, the forest habitat of...fruitbats...in Southeast Asia has been substantially reduced by deforestation for pulpwood and industrial plantation...[A series of weather events then] led to acute reduction in the availability of flowering and fruiting forest trees for foraging by flying-foxes in their already shrinking habitat...This culminated in an unprecedented encroachment of flying-foxes into cultivated fruit orchards in the initial outbreak area...These anthropogenic events, coupled with the location of piggeries in orchards and the design of pigsties in the index farms allowed transmission of a novel paramyxovirus from its reservoir host to the domestic pig and ultimately to the human population and other domestic animals.

We argue in contrast, by their diversity in time, space and mode, traditional and conservation agricultures can create numerous functional barriers—a kind of sterilizing temperature—limiting pathogen evolution and spread (Diaz et al. 2006; Burdon and Thrall 2008; Perfecto and Vandermeer 2010; R. Wallace and R.G. Wallace 2015).

For instance, pastoralist livestock production, until recently supplying 67 % of Kenya's available meat, promotes wildlife conservation in the semi-arid Sub-Saharan (Kock 2005; Kock et al. 2010). The last of the unrestricted ungulate populations are associated with such systems. Disease plays a critical role for both semi-arid wildlife and livestock, including indirectly multiplying the impact of other pressures such as predation. Foot-and-mouth disease and malignant catarrhal fever, among other diseases, selected for strict limits on land use and animal movement.

But growing awareness of the complexities of the wildlife-livestock interface in association with a management philosophy erring on the side of integration have revised thinking around disease control in rangelands. First, there appears little actual physical contact between livestock and wildlife, even in integrated semi-arid systems. The separation is driven by the animals' own niche behavior, leaving disease spillover between wild and domesticated animals largely sapronotic and vector-borne (e.g., anthrax, Rift Valley fever, and theileriosis). Accumulating molecular evidence indicates that many of the diseases wildlife and livestock putatively transmit to each other are in fact host-specific products of vicariant cospeciation, although host switching does intermittently occur (Huyse et al. 2005; Foster et al. 2009; Widmer and Akiyoshi 2010; Ruecker et al. 2012; Hoberg and Brooks 2015).

Diseases that have had transcontinental impact (e.g., rinderpest, bovine tuberculosis) have been imported from abroad and are density-dependent, part and parcel of millenia of anthropogenic disruption (Kock et al. 2010). In these cases, vaccination and fencing have been deployed. Control has also now turned to separating livestock into small "export zones" in which livestock are fattened and from which wildlife are excluded, synergistically supporting pastoralist pluriactivity. Behavioral adaptations are also introduced by the pastoralist themselves. Kock (2005) describes one such adaptation controlling malignant catarrhal fever:

> Generally, the livestock keepers will avoid calving grounds, but when they have no choice the grazing strategy of the pastoralists shows considerable understanding of the epidemiology of the disease. The virus is highly sensitive to drying, heat, and ultraviolet light, so under natural conditions the pastoralists have learned that since wildebeest calve at night, by 10:00 a.m. the pasture is sterile (in terms of MCF) and infection can be avoided.

The economies of scale of capitalized production remove such functional diversity and are paradoxically extended to those pathogens its agrosystems select, even in systems exercising model biocontrol (Brown 2004; R.G. Wallace 2009; Hogerwerf et al. 2010).

For Ebola, the 2008 REBOV outbreak and the transmission studies of deadly ZEBOV to and between pigs together imply commoditized husbandry is open to devolving into a "plague plantations" across a large part of the world, particularly in the USA, Europe, and China, but increasingly other states, as the Philippines demonstrates, where intensive confined feedlots have come to dominate animal protein production. Commodity hog is producing industrial densities in growing areas of West and East Africa alike, including in Guinea-Bissau, Burkina Faso, Togo, Nigeria, Rwanda, and Uganda, several of which have hosted Ebola outbreaks (Atherstone et al. 2014) (Fig. 2.4b). Ebola outbreaks in Uganda have taken place in areas with high pig density (Weingartl et al. 2013). We hypothesize REBOV's transition in the Philippines foreshadows the kind of agroeconomic domain shift Ebola may already be undertaking in Africa.

One need be concerned about whatever hog mortality may result, of course, with clear implications for human health (Bausch 2011). But subclinical infections are themselves emblematic of a growing prevalence, especially with REBOV now in China, where half the world's hog population is raised. The greater the population infected, the larger the "laboratory" in which Ebola can recombine, coinfect with other prevalent industrial pathogens, and experiment with human-specific phenotypes (Frank 1994; R.G. Wallace 2009; Mennerat et al. 2010; Drew 2011; FAO 2013b; Fuller et al. 2013).

Contrary to the sector's operational premises, pathogens evolve, or, by shifting host population states alone, are maneuvered into new evolutionary stable strategies (Levin 1968; Lipsitch and Nowak 1995; R.G. Wallace et al. 2014). Hog's expanding economic geographies may produce a "critical connectivity" across farms at multiple spatial scales that changes the rules under which the evolution of virulence takes place, permitting virulent strains previously selected against by a cap on available susceptibles to invade local variants (Boots et al. 2004; Messinger and Ostling 2009; Atkins et al. 2010).

The damage imposed by such economies of scale is routinely described in the industrial literature, with considerable attention given to the means by which to offload associated costs (e.g., Arthur and Albers 2003; Fulton 2006). By a moral hazard of existential significance, the overhead is externalized to the occupational hazards of production; livestock health and welfare; wildlife by way of deforestation, pollution and disease spillover; local environs polluted by manure runoff; consumer health via food-borne pathogens and metabolic disorders; and damage to transportation and health infrastructure (Singer 2005; Catelo et al. 2008; R.G. Wallace and Kock 2012; Leonard 2014; Genoways 2014). Such costs are routinely floated by taxpayers and governance across administrative units. Subsidies and bailouts benefit individual companies exercising political influence but are also deployed as a matter of national economic policy (R.G. Wallace 2009, RG Wallace 2012).

Bello (2003), for instance, describes a gambit by which the very efforts to reverse Filipino protectionism were parlayed into a means by which to block outbound domestic exports on epizootic grounds:

> Even Australia, an ally of the Philippines in the so-called Cairns Group, a grouping of developed and developing agro-exporting countries, beat up on the Philippines by invoking sanitary and phytosanitary standards, a standard Washington tactic. In mid-2002, after years of being petitioned to admit Philippine cavendish bananas, the Australian government decided against the import. The ostensible reason was the risk of the Philippine banana carrying pests and diseases that could ruin the Australian banana industry. Yet the Philippine bananas had been shipped since the sixties to countries with high quarantine standards, including Japan and New Zealand.

Much of the academic and nongovernmental response to the disease challenges such production raises appears two-pronged in nature (R.G. Wallace et al. 2015). The first approach is a multilevel analysis described as broad in scope that also omits the ideological commitments behind the shifts in land use driving emergent disease. Plowright et al. (2015), for instance, layer the causes underlying the emergence of bat zoonotics from virion and immune response up through community ecology and anthropogenic context. Yet such a comprehensive analysis includes a telling omission:

> Coincident with [expanding urban and periurban land use and increasing presence of horses] is another [factor]: the range of black flying-fox is expanding rapidly southward at rates faster than projected on the basis of climate change scenarios...Black flying-fox have a stronger association with Hendra virus spillover events than other flying-fox species..., and may be more likely to feed on the marginal foods that support resident populations in

anthropogenic landscapes…Thus, the range shift of black flying-fox may contribute to the increasing incidence and recent southern extension of Hendra virus spillover events in the subtropics.

The underlying neoliberal shifts in land markets, agricultural consolidation, and biosecurity that are driving bat dynamics in the Hendra zone are left unremarked upon (McKenzie 2011; Beer 2012; Higgins et al. 2012; Maye et al. 2012, and Lawrence et al. 2013). Plowright et al. avoid even "deforestation" in spite of the evidence, including accumulated elsewhere by several co-authors, linking forest loss to bat dynamics and Hendra (Field et al. 2001; Daszak et al. 2006; McFarlane et al. 2011).

The second archetypal response gives the impression of convergent accommodation (e.g., Khan et al. 2013). One Health proponents Morse et al. (2013), advocating wildlife, livestock, and human health be simultaneously addressed, appear to frame intervention around *assisting* such development:

> The challenge to true pandemic prevention (and pre-emption) is how to address the underlying drivers that are essentially ecological (e.g., juxtaposition of livestock production and wildlife populations) or occur on large spatial scales because of economic activity (e.g., change in land use related to development of tropical forests)…Incentives for industries with roles in activities that propagate pandemics could be linked to development initiatives. For example, concessions in development of logging or mining could include better food supply chains as an alternative to bushmeat hunting, better clinics for migrant workers than are available, and more intensive surveillance of livestock at these crucial interfaces… [D]evelopment agencies [are now focusing] on improving individual countries' abilities to identify new zoonoses early and mitigate quickly any new health threats arising within their borders.

There is a grave irony in replacing bushmeat with the better food supply chains Morse et al. recommend. Clearly de-commoditizing bushmeat associated with logging and mining camps is a step forward in disease control (Bowen-Jones et al. 2003), but should intensive agriculture take its stead, forest pathogens may now regularly access human populations, accelerating the very processes that appear to have led to the ZEBOV outbreak in West Africa.

No one need oppose better clinics for migrant workers to conclude Morse et al. lay individualistic responsibility for cleaning up after such "drivers" on the smallholders and indigenous populations neoliberal agriculture expropriates:

> The importance of human exposure throughout the [pandemic] emergence process also suggests that simple behavioural precautions could greatly reduce risk. Risks to hunters, food handlers, and livestock workers from occupational exposure could be reduced in hotspots of emerging infectious diseases through routine sanitation and biosafety precautions…

The "hotspots of emerging infectious diseases" now span a vast and synergistic amalgam of expanding commodity plantation and confined feedlot livestock and poultry. Such a system, reducing the ecosystemic "friction" against pathogen emergence below many a region's controlling threshold, requires a level of biosecure containment beyond the economic margins and practical limits of its own model of production. In this context, advocating washing hands plays as disingenuous political theater, particularly in countries where millions live without access to clean water and much of what is available is pledged to commodity agriculture.

In contrast, a Structural One Health offers an approach that, among other possibilities, explicitly addresses the relationships among transnational circuits of capital, shifts in agroecological landscapes, and the emergence of new diseases (R.G. Wallace et al. 2015). Such foundational context, extending well beyond an epicenter's borders, offers a critical entry into devising successful interventions into human pathogens originating out of agriculture.

Acknowledgements The authors thank Marius Gilbert and Thomas Van Boeckel for perspicacious comment. The research reported here is part of a line of research organized as the Ebola Agroeconomic Systems Team (EAST). Partial support for this publication came from a Eunice Kennedy Shriver National Institute of Child Health and Human Development research infrastructure grant, R24 HD042828, to the Center for Studies in Demography and Ecology at the University of Washington.

Mathematical Appendix

Epidemic Prevention Farming

The models above focus primarily on explosive epidemic outbreaks and their containment costs, incorporating as well the influence of sudden Levy jumps. In general, however, endemic levels of infection, fluctuating about some mean, would be expected, and the central question then surrounds the transition between endemic and epidemic modes.

Khasminskii's (1966/2006, 2012, Theorem 4.1) version of Eq. (2.1) provides insight, using a linear first approximation to some complicated, multidimensional, cross-influence function expanded about a quasi-stable equilibrium point. This gives the system of stochastic equations

$$dx_t^i = \sum_{j=1}^{l} b_i^j x_t^j dt + \sum_{r=1}^{n} \sum_{j=1}^{l} \sigma_{i,r}^j x_t^j dW_t^r \quad i = 1, \ldots, l \qquad (2.21)$$

where dW_t^r is white noise and the b and σ terms are constants.

Khasminskii defines two associated matrices,

$$a_{i,j}(x) = \sum_{k,s=1}^{l} \sum_{r=1}^{n} \sigma_{i,r}^k \sigma_{j,r}^s x^k x^s, \quad \mathbf{B} = ||b_i^j||$$

under the condition that, for **A**, the inner product condition

$$(\mathbf{A}(x)\alpha, \alpha) \geq m|x|^2 |\alpha|^2 \qquad (2.22)$$

always holds.

Khasminskii invokes two new variates, $\lambda = x/|x|$ on the unit sphere, and $\rho = \log[|x|]$, expanding $d\rho_t$ using the Ito chain rule to obtain

$$
d\rho_t = \left[(\mathbf{B}\lambda_t, \lambda_t) + \frac{1}{2} \sum_{i=1}^{l} a_{i,i}(\lambda_t) - \sum_{i,j=1}^{l} a_{i,j}(\lambda_t)\lambda_t^i \lambda_t^j \right] dt
$$
$$
+ \sum_r (\sigma(r)\lambda_t, \lambda_t) dW_t^r \tag{2.23}
$$

where $\sigma(r) = ||\sigma_{i,r}^j||, i,j = 1,\ldots,l.$

Define

$$
Q(\lambda) = (\mathbf{B}\lambda, \lambda) + \frac{1}{2} \sum_{i=1}^{l} a_{i,i}(\lambda) - \sum_{i,j=1}^{l} a_{i,j}(\lambda)\lambda^i \lambda^j
$$
$$
J = \int Q(\lambda)d\lambda
$$

where the integral is taken over the unit sphere. (Khasminskii 1966/2006, 2012, Theorem 4.1) shows that, if $J < 0$, the complex stochastic process converges to an endemic equilibrium distribution. If $J > 0$, then the probability that $|x_t| \to \infty$ as $t \to \infty$ is 1.

Thus, for any given cross-influence matrix $\mathbf{B}$, there is a set of structures defined by the matrix $\mathbf{A}$—under the condition of Eq. (2.22)—that will contain a pathogen outbreak to endemic levels. Conversely, given an endemic distribution, sufficient alteration of either the structural matrix $\mathbf{B}$ or of the "noise" matrix $\mathbf{A}$ would trigger an epidemic outbreak $|x_t| \to \infty$.

Extension of this result involving jump processes can be found in Khasminskii et al. (2007).

Large Deviations

Something similar to Eq. (2.14) can be simply derived via a standard large deviations argument.

Following Dembo and Zeitouni (1998), let $X_1, X_2, \ldots X_n$ be a sequence of independent, standard Normal, real-valued random variables and let

$$
S_n = \frac{1}{n} \sum_{j=1}^{n} X_j \tag{2.24}
$$

Since S_n is again a Normal random variable with zero mean and variance $1/n$, for all $\delta > 0$

$$\lim_{n \to \infty} P(|S_n| \geq \delta) = 0 \tag{2.25}$$

where P is the probability that the absolute value of S_n is greater or equal to δ. Some manipulation, however, gives

$$P(|S_n| \geq \delta) = 1 - \frac{1}{\sqrt{2\pi}} \int_{-\delta\sqrt{n}}^{\delta\sqrt{n}} \exp(-x^2/2)dx \tag{2.26}$$

so that

$$\lim_{n \to \infty} \frac{\log P(|S_n| \geq \delta)}{n} = -\delta^2/2 \tag{2.27}$$

This can be rewritten for large n as

$$P(|S_n| \geq \delta) \approx \exp(-n\delta^2/2) \tag{2.28}$$

That is, for large n, the probability of a large deviation in S_n follows something much like the asymptotic equipartition relation of the Shannon–McMillan Theorem.

This result can be generalized to more complicated probability spaces using Sanov's Theorem, the Gartner–Ellis Theorem, and related developments (Dembo and Zeitouni 1998) to show that large deviations paths of length n all have approximately the probability

$$P(n) \propto \exp(-n\mathcal{H}[\mathbf{X}]) \tag{2.29}$$

where $\mathcal{H}$ is of the form $-\sum_i P_i \log(P_i)$ for some probability distribution. Under the conditions of our analysis, $P(n)$ is the probability of an excursion from the absorbing state of $n =$ zero infections. $\mathcal{H}$ thus quantifies an information source representing the active imposition of control strategies to prevent a large-scale outbreak of infection, and the Black–Scholes cost analysis carries through.

References

Alawneh J. I., Barnes, T. S., Parke, C., Lapuz, E., David, E., Basinang, V., et al. (2014). Description of the pig production systems, biosecurity practices and herd health providers in two provinces with high swine density in the Philippines. *Preventive Veterinary Medicine, 114*(2), 73–87. doi:10.1016/j.prevetmed.2014.01.020. Epub 2014 Jan 29.

Alexandratos, N., & Bruinsma, J. (2012). *World Agriculture Towards 2030/2050: The 2012 Revision.* ESA Working Paper 12-03. Available online at http://www.fao.org/docrep/016/ap106e/ap106e.pdf.

Allen, J., & Lavau, S. (2014). Just-in-time disease: Biosecurity, poultry and power. *Journal of Cultural Economics*. http://dx.doi.org/10.1080/17530350.2014.904243.

Arthur, J. A., & Albers, G. A. (2003). Industrial perspective on problems and issues associated with poultry breeding. In: W. M. Muir, & S. E. Aggrey (Eds.), *Poultry genetics, breeding and biotechnology*. Oxfordshire: CABI Publishing.

Atherstone, C., Roesel, K., & Grace, D. (2014). *Ebola risk assessment in the pig value chain in Uganda*. ILRI Research Report 34, International Livestock Research Institute, Nairobi, Kenya.

Atkins, K., Wallace, R. G., Hogerwerf, L., Gilbert, M., Slingenbergh, J., Otte, J., et al. (2010). *Livestock landscapes and the evolution of influenza virulence*. Virulence Team Working Paper No. 1, Animal Health and Production Division, Food and Agriculture Organization of the United Nations, Rome.

Bailey, N. T. (1975). *The mathematical theory of infectious diseases and its applications*. New York: Hafner Press.

Balk, D. L., Deichmann, U., Yetman, G., Pozzi, F., Hay, S. I., & Nelson, A. (2006). Determining global population distribution: Methods, applications and data. *Advances in Parasitology, 62*, 119–156. http://dx.doi.org/10.1016.

Baron, R. C., McCormick, J. B., & Zubeir, O. A. (1983). Ebola virus disease in southern Sudan: Hospital dissemination and intrafamilial spread. *Bulletin of the World Health Organization, 61*, 99–1003.

Barrette, R. W., Metwally, S. A., Rowland, J. M., Xu, L., Zaki, S. R., Nichol, S. T., et al. (2009). Discovery of swine as a host for the Reston ebolavirus. *Science, 325*(5937), 204–206. doi:10.1126/science.1172705.

Bartsch, S. M., Gorham, K., & Lee, B. Y. (2015). The cost of an Ebola case. *Pathogens and global health, 109*(1), 4–9.

Bausch, D. (2011). Ebola virus as a foodborne pathogen? Cause for consideration but not panic. *Journal of Infectious Disease, 204*, 179–181.

Bausch, D., & Schwarz, L. (2014). Outbreak of Ebola virus disease in Guinea: Where ecology meets economy. *PLOS Neglected Tropical Diseases, 8*, e3056.

Beddington, J., & May, R. (1977). Harvesting natural populations in a randomly fluctuating environment. *Science, 197*, 463–465.

Beer, A. (2012). The economic g of Australia and its analysis: From industrial to post-industrial regions. *Geographical Research, 50*(3), 269–281.

Bello, W. (2003). *Multilateral punishment: The Philippines in the WTO, 1995–2003*. Stop the New Round Coalition! Manila: Focus on the Global South. http://www.apl.org.ph/ps/multilateral-punishment.pdf.

Bergmann, L. R. (2013a). Bound by chains of carbon: Ecological-economic geographies of globalization. *Annals of the Association of American Geographers, 103*, 1348–1370. http://dx.doi.org/10.1080/00045608.2013.779547.

Bergmann, L. R. (2013b). *Beyond the Anthropocene: Toward modest mathematical narratives for more-than-human global communities*. Paper accepted for session: 'Re-evaluating the Anthropocene, Resituating Anthropos,' Annual Meeting of the Association of American Geographers, Los Angeles.

Bertherat, E., Renaut, A., Nabias, R., Dubreuil, G., & Georges-Courbot, M. C. (1999). Leptospirosis and Ebola virus infection in five gold-panning villages in northeastern Gabon. *The American Journal of Tropical Medicine and Hygiene, 60*(4), 610–615.

Black, F., & Scholes, M. (1973). The pricing of options and corporate liabilities. *Journal of Political Economy, 81*, 637–654.

Boni, M. F., Galvani, A. P., Wickelgrend, A. L., & Malani, A. (2013). Economic epidemiology of avian influenza on smallholder poultry farms. *Theoretical Population Biology, 90*, 135e144.

Borras, S. M., & Franco, J. C. (2011). *Political dynamics of land-grabbing in Southeast Asia: Understanding Europe's role*. Amsterdam: Transnational Institute.

Bowen-Jones, E., Brown, D., & Robinson, E. J. Z. (2003). Economic commodity or environmental crisis? An interdisciplinary approach to analysing the bushmeat trade in Central and West Africa. *Area, 35*(4), 390–402.

Boyd, I. L., Freer-Smith, P. H., Gilligan, C. A., & Godfray, H. C. (2013). The consequence of tree pests and diseases for ecosystem services. *Science, 342*, 1235773. doi:10.1126/science.1235773.

Brown, C. (2004). Emerging zoonoses and pathogens of public health significance – an overview. *Scientific and Technical Review of the Office International des Epizooties (Paris), 23*(2), 435–442.

Burdon, J., & Thrall, P. (2008). Pathogen evolution across the agro-ecological interface: Implications for management. *Evolutionary Applications, 1*, 57–65.

Cai, Y., Wang, X., Wang, W., & Zhao, M. (2013). Stochastic dynamics of an SIRS epidemic model with ratio-dependent incidence rate. *Abstract and Applied Analysis*, 11pp. ID 172631.

Capua, I., & Marangon, S. (2007). Control and prevention of avian influenza in an evolving scenario. *Vaccine, 25*(30), 5645–5652.

Cardinoza, G., & Reyes, C. (2009, January 8). UN agencies inspect Luzon hog farms. *Philippine Daily Inquirer*. http://newsinfo.inquirer.net/breakingnews/nation/view/20090108-182214/UN-agencies-inspect-Luzon-hog-farms.

Carroll, A., Towner, J. S., Sealy, T. K., McMullan, L. K., Khristova, M. L., Burt, F. J., et al. (2013). Molecular evolution of viruses of the family *Filoviridae* based on 97 whole-genome sequences. *Journal of Virology, 87*, 2608–2616.

Catelo, M. A. O., Narrod, C. A., & Tiongco, M. M. (2008). *Structural Changes in the Philippine Pig Industry and Their Environmental Implications*. IFPRI Discussion Paper 00781, The International Food Policy Research Institute, Washington, DC.

Centeno, M. A., & Cohen, J. N. (2012). The arc of Neoliberalism. *Annual Review of Sociology, 38*, 317–340. doi:10.1146/annurev-soc-081309-150235.

Chua, K. B. (2003). Nipah virus outbreak in Malaysia. *Journal of Clinical Virology, 26*, 265–275.

Costales, A., Delgado, C., Catelo, M. A., Lapar, M. L., Tiongco, M., Ehui, S., et al. (2007). *Scale and access issues affecting smallholder hog producers in and expanding peri-urban market; Southern Luzon, Philippines*. IFPRI Research Report No. 151. IFPRI, Washington, DC. http://www.ifpri.org/sites/default/files/publications/rr151_0.pdf.

Costales, A. C., Delgado, C., Catelo, M. A. O., Tiongco, M., Chatterjee, A., delos Reyes, A., & Narrod, C. (2003). *Policy, technical, and environmental determinants and implications of the scaling-up of broiler and swine production in The Philippines*. Annex I, Final Report of IFPRI-FAO Livestock Industrialization Project: Phase, I. I., International Food Policy Research Institute, Washington, DC.

Coxhead, I., & Jayasuriya, S. (2002). *Development Strategy, Poverty and Deforestation in the Philippines*. Department of Agricultural & Applied Economics, University of Wisconsin-Madison, Staff Paper No. 456. http://www.aae.wisc.edu/pubs/sps/pdf/stpap456.pdf.

Coxhead, I., Shively, G., & Shuai, X. (1999). *Development Policies, Resource Constraints, and Agricultural Expansion on the Philippine Land Frontier*. Department of Agricultural & Applied Economics, University of Wisconsin-Madison, Staff Paper No. 425. https://www.aae.wisc.edu/pubs/sps/pdf/stpap425.pdf.

Cyranoski, D. (2009). Ebola outbreak has experts rooting for answers. *Nature, 457*, 364–365.

Daszak, P., Plowright, R. K., Epstein, J. H., Pulliam, J., Abdul Rahman, S., Field, H. E., et al. (2006). The emergence of Nipah and Hendra virus: Pathogen dynamics across a wildlife-livestock-human continuum. In S. K. Collinge & Ray, C. (Eds.), *Disease ecology: Community structure and pathogen dynamics* (pp. 186–201). Oxford: Oxford University Press.

David, C. C. (1997). *Agricultural policy and the WTO agreement: The Philippine case*. Discussion Paper Series No. 97–13, Philippine Institute for Development Studies, Makati City. http://dirp4.pids.gov.ph/ris/dps/pidsdps9713.pdf.

Delgado, C. L., Narrod, C., & Tiongco, M. M. (2003). *Policy, technical, and environmental determinants and implications of the scaling-up of livestock production in four fast-growing developing countries: A synthesis*. Final Research Report of Phase, I. I., Project on Livestock Industrialization, Trade and Social-Health-Environment Impacts in Developing Countries. http://www.fao.org/3/a-x6170e/x6170e00.htm.

Dembo, A., & Zeitouni, O. (1998). *Large deviations: Techniques and applications*. New York: Springer.

Diaz, S., Fargione, J., Chapin, F. S., & Tilman, D. (2006). Biodiversity loss threatens human well-being. *PLoS Biology, 4*(8), e277. doi:10.1371/journal.pbio.0040277.

Drew, T. W. (2011). The emergence and evolution of swine viral diseases: To what extent have husbandry systems and global trade contributed to their distribution and diversity? *Revue Scientifique et Technique (Paris), 30*, 95–106.

Dudas, G., & Rambaut, A. (2014). Phylogenetic analysis of Guinea 2014 EBOV Ebolavirus outbreak 2014. *PLOS Currents Outbreaks, 6.* ecurrents.outbreaks.84eefe5ce43ec9dc0bf0670f7b8b417d.

Engering, A., Hogerwerf, L., & Slingenbergh, J. (2013). Pathogen host environment interplay and disease emergence. *Emerging Microbes and Infections, 2*, e5. http://www.nature.com/emi/journal/v2/n2/full/emi20135a.htm.

Epstein, J. H., Field, H. E., Luby, S., Pulliam, J. R., & Daszak, P. (2006). Nipah virus: Impact, origins, and causes of emergence. *Current Infectious Disease Reports, 8*(1), 59–65.

Ercsey-Ravasz, M., Toroczkai, Z., Lakner, Z., & Baranyi, J. (2012). Complexity of the international agro-food trade network and its impact on food safety. *PLoS One, 7*(5), e37810. http://dx.doi.org/10.1371/journal.pone.0037810.

FAO (2013a). *FAO Statistical Yearbook 2013*. Food and Agriculture Organization, United Nations, Rome.

FAO (2013b). *World Livestock 2013: Changing disease landscapes*. Food and Agriculture Organization, United Nations, Rome.

Farber, S. C., Costanza, R., & Wilson, M. A. (2002). Economic and ecological concepts for valuing ecosystem services. *Ecological Economics, 41*, 375–392.

Field, H., Young, P., Yob, J. M., Mills, J., Hall, L., & Mackenzie, J. (2001). The natural history of Hendra and Nipah viruses. *Microbes and Infection, 3*(4), 307–314.

Foley, J., Defries, R., Asner, G. P., Barford, C., Bonan, G., Carpenter, S. R., et al. (2005). Global consequences of land use. *Science, 309*, 570–574.

Formenty, P., Boesch, C., Wyers, M., Steiner, C., Donati, F., Dind, F., et al. (1999). Ebola virus outbreak among wild chimpanzees living in a rain forest of Cote d'Ivoire. *The Journal of Infectious Diseases, 179*(1), S120–126.

Foster, J. T., Beckstrom-Sternberg, S. M., Pearson, T., Beckstrom-Sternberg, J. S., Chain, P. S., Roberto, F. F., et al. (2009). Whole-genome-based phylogeny and divergence of the genus Brucella. *Journal of Bacteriology, 191*(8), 2864–2870. doi:10.1128/JB.01581-08.

Fuller, T. L., Gilbert, M., Martin, V., Cappelle, J., Hosseini, P., Njabo, K. Y., et al. (2013). Predicting hotspots for influenza virus reassortment. *Emerging Infectious Diseases, 19*(4), 581–588. doi:10.3201/eid1904.120903.

Fulton, J. E. (2006). Avian genetic stock preservation: An industry perspective. *Poultry Science, 85*(2), 227–231. Paper for the Poultry Science Association Ancillary Scientists Symposium, July 31, 2005, Auburn, Alabama, "Conservation of Avian Genetic Resources: Current Opportunities and Challenges," organized and chaired by Dr. Muquarrab Qureshi.

Ganti, T. (2014). Neoliberalism. *Annual Review of Anthropology, 43*, 89–104. doi:10.1146/annurev-anthro-092412-155528.

Gatherer, D. (2015). The unprecedented scale of the West African Ebola virus disease outbreak is due to environmental and sociological factors, not special attributes of the currently circulating strain of the virus. *Evidence-Based Medicine, 20*(1), 28. doi:10.1136/ebmed-2014-110127.

Genoways, T. (2014). *The Chain: Farm, Factory, and the Fate of Our Food*. New York: Harper-Collins.

Gerber, P. J., Steinfeld, H., Henderson, B., Mottet, A., Opio, C., Dijkman, J., et al. (2013). *Tackling Climate Change Through Livestock – A Global Assessment of Emissions and Mitigation Opportunities*. Food and Agriculture Organization of the United Nations (FAO), Rome. http://www.fao.org/docrep/018/i3437e/i3437e.pdf.

Gilbert, M., & Pfeiffer, D. U. (2012). Risk factor modelling of the spatio-temporal patterns of highly pathogenic avian influenza (HPAIV) H5N1: A review. *Spatial and spatio-temporal epidemiology, 3*(3), 173–183. doi:10.1016/j.sste.2012.01.002.

Gire, S. K., Goba, A., & Andersen, K. G. (2014). Genomic surveillance elucidates Ebola virus origin and transmission during the 2014 outbreak. *Science, 345,* 1369–1372.

Gonzalez, J.-P., Herbreteau, V., Morvan, J., & Leory, E. (2005). Ebola virus circulation in Africa: A balance between clinical expression and epidemiological silence. *Bulletin de la Societe de pathologie exotique, 98*(3), 210–221.

Gould, P., & Wallace, R. (1994). Spatial structures and scientific paradoxes in the AIDS pandemic. *Geofrafiska Annaler, 76B,* 105–116.

Graham, J. P., Leibler, J. H., Price, L. B., Otte, J. M., Pfeiffer, D. U., Tiensin, T., et al. (2008). The animal-human interface and infectious disease in industrial food animal production: Rethinking biosecurity and biocontainment. *Public Health Reports, 123,* 282–299.

Groseth, A., Feldmann, H., & Strong, J. E. (2007). The ecology of Ebola virus. *Trends in Microbiology, 15*(9), 408–416.

Guo, H. G., Pati, S., Sadowska, M., Charurat, M., & Reitz, M. (2004). Tumorigenesis by human herpesvirus 8 vGPCR is accelerated by human immunodeficiency virus type 1 Tat. *Journal of Virology, 78*(17), 9336–9342.

Habito, C. F. (2011). Economy, environment and Filipino children. In F. Rosario-Braid, R. R. Tuazon, & Lopez, A. L. C. (Eds.), *The future of Filipino children: Development issues and trends.* UNICEF and Asian Institute of Journalism and Communication. http://www.aijc.com.ph/Megatrend%20final%20complete.pdf.

Halpin, K., & Mungall, B. (2007). Recent progress in henipavirus research. *Comparative Immunology, Microbiology and Infectious Diseases, 30,* 287–307.

Harvey, D. (2005). *A brief history of Neoliberalism.* Oxford: Oxford University Press.

Hayman, D. T., Bowen, R. A., Cryan, P. M., McCracken, G. F., O'Shea, T. J., Peel, A. J., et al. (2013). Ecology of zoonotic infectious diseases in bats: Current knowledge and future directions. *Zoonoses Public Health, 60*(1), 2–21. doi:10.1111/zph.12000.

Heaney, L. R., Walsh, J. S. Jr., & Peterson, A. T. (2005). The roles of geological history and colonization abilities in genetic differentiation between mammalian populations in the Philippine archipelago. *Journal of Biogeography, 32,* 229–247.

Herrero, M., Havlík, P., Valin, H., Notenbaert, A., Rufino, M. C., Thornton, P. K., et al. (2013). Biomass use, production, feed efficiencies, and greenhouse gas emissions from global livestock systems. *Proceedings of the National Academy of Sciences of the United States of America, 110*(52), 20888–20893. doi:10.1073/pnas.1308149110.

Higgins, V., Dibden, J., & Cocklin, C. (2012). Market instruments and the neoliberalisation of land management in rural Australia. *Geoforum, 43,* 377–386.

Hinchliffe, S. (2015). More than one world, more than one health: Re-configuring interspecies health. *Social Science & Medicine, 129,* 28–35.

Hoberg, E. P., & Brooks, D. R. (2015). Evolution in action: Climate change, biodiversity dynamics and emerging infectious disease. *Philosophical Transactions B, 370,* 20130553.

Hogerwerf, L., Houben, R., Hall, K., Gilbert, M., Slingenbergh, J., & Wallace, R. G. (2010). *Agroecological resilience and protopandemic influenza.* Final report, Animal Health and Production Division, Food and Agriculture Organization, Rome.

Horsthemeke, W., & Lefever, R. (2006). *Noise-induced transitions* (Vol. 15). Theory and applications in physics, chemistry, and biology. New York: Springer.

Humphries-Waa, K., Drake, T., Huszar, A., Liverani, M., Borin, K., Touch, S., et al. (2013). Human H5N1 influenza infections in Cambodia 2005–2011: Case series and cost-of-illness. *BMC Public Health, 13,* 549. doi:10.1186/1471-2458-13-549.

Huyse, T., Poulin, R., & Theron, A. (2005). Speciation in parasites: A population genetics approach. *Trends in Parasitology, 21*(10), 469–475.

Ilinykh, P. A., Lubaki, N. M., Widen, S. G., Renn, L. A., Theisen, T. C., Rabin, R. L., et al. (2015). Different temporal effects of Ebola virus VP35 and VP24 proteins on the global gene expression in human dendritic cells. *Journal of Virology,* 00924-15.doi:10.1128/JVI.00924-15.

Jones, K. E., Mickleburgh, S. P., Sechrest, W., & Walsh, A. L. (2009). Global overview of the conservation of island bats: Importance, challenges, and opportunities. In T.H. Fleming, & P.A. Racey (Eds.), *Island bats: Evolution, ecology, and conservation*. Chicago: University of Chicago Press.

Jones, B. A., Grace, D., Kock, R., Alonso, S., Rushton, J., Said, M. Y., et al. (2013). Zoonosis emergence linked to agricultural intensification and environmental change. *PNAS, 110*, 8399–8404.

Jones, J. (2011). Ebola, emerging: The limitations of culturalist discourses in epidemiology. *The Journal of Global Health, 1*, 1–6. http://www.ghjournal.org/?p=6194.

Kelly, P. F. (1998). The politics of urban-rural relations: Land use conversion in the Philippines. *Environment and Urbanization, 10*, 35–54.

Kelly, P. F. (2000). *Landscapes of globalization: Human geographies of economic change in the Philippines*. London: Routledge.

Kelly, P. F. (2011). Migration, agrarian transition, and rural change in Southeast Asia. *Critical Asian Studies, 43*(4), 479–506. doi:10.1080/14672715.2011.623516.

Kemeny, J., & Snell, J. (1976). *Finite Markov Chains*. New York: Springer.

Kentikelenis, A., King, L., McKee, M., & Stuckler, D. (2014). The international monetary fund and the ebola outbreak. *The Lancet Global Health*. http://dx.doi.org/10.1016/S2214-109X(14)70377-8.

Khan, S. U., Atanasova, K. R., Krueger, W. S., Ramirez, A., & Gray, G. C. (2013). Epidemiology, geographical distribution, and economic consequences of swine zoonoses: A narrative review. *Emerging Microbes & Infections, 2*, e92. doi:10.1038/emi.2013.87.

Khasminskii, R. (1966). Necessary and sufficient conditions for the asymptotic stability of linear stochastic systems. *Theory of Probability and its Applications, 12*, 144–147. (In Russian, translated by B. Seckler, 2006).

Khasminskii, R. (2012). *Stochastic stability of differential equations*. New York: Springer.

Khasminskii, R., Zhu, C., & Yin, G. (2007). Stability of regime-switching diffusions. *Stochastic Processes and their Applications, 117*, 1037–1051.

Kleczkowski, A., Oleś, K., Gudowska-Nowak, E., & Gilligan, C. A. (2012). Searching for the most cost effective strategy for controlling epidemics spreading on regular and small-world networks. *Journal of the Royal Society, Interface, 9*(66), 158–169. doi:10.1098/rsif.2011.0216.

Kobinger, G., Leung, A., Neufeld, J., Richardson, J. S., Falzarano, D., Smith, G., et al. (2011). Replication, pathogenicity, shedding, and transmission of Zaire ebolavirus in pigs. *Journal of Infectious Disease, 204*, 200–208.

Kock, R. A. (2005). What is this infamous 'Wildlife/Livestock Disease Interface'? A review of current knowledge for the African continent. In S. A. Osofsky, S. Cleaveland, W. B. Karesh, M. D. Kock, P. J. Nyhus, L. Starr & A. Yang, A. (Eds.), *Conservation and development interventions at the wildlife/livestock interface implications for wildlife, livestock and human health* (pp. 1–13). Gland, and Cambridge: IUCN.

Kock, R. A., Alders, R., & Wallace, R. G. (2012). Wildlife, wild food, food security and human society. In *Animal health and biodiversity - preparing for the future. Illustrating contributions to public health* (pp. 71e79). Compendium of the OIE Global Conference on Wildlife 23–25 February 2011 Paris.

Kock, R., Kock, M., Cleaveland, S., & Thomson, G. (2010). Health and disease in wild rangelands. In J. du Toit, R. Kock, & J. Deutsch (Eds.), *Wild rangelands: Conserving wildlife while maintaining livestock in semi-arid ecosystems*. Oxford: Wiley-Blackwell.

Knight-Jones, T. J., & Rushton, J. (2013). The economic impacts of foot and mouth disease - what are they, how big are they and where do they occur? *Preventive Veterinary Medicine, 112*(3–4), 161–73. doi:10.1016/j.prevetmed.2013.07.013.

Lapus, Z. M. (2014). Pork outlook 2014: Philippines. What the experts say. pig333.com. https://www.pig333.com/what-the-experts-say/pork-outlook-2014-philippines-8158/.

Lawn, S. D. (2004). AIDS in Africa: The impact of coinfections on the pathogenesis of HIV-1 infection. *Journal of Infection, 48*, 1–12.

Lawrence, G., Richards, C., & Lyons, K. (2013). Food security in Australia in an era of neoliberalism, productivism and climate change. *Journal of Rural Studies, 29*, 30–39.

Lehrnbecher, T. L., Foster, C. B., Zhu, S., Venzon, D., Steinberg, S. M., Wyvill, K., et al. (2000). Variant genotypes of FcgammaRIIIA influence the development of Kaposi's sarcoma in HIV-infected men. *Blood, 95*(7), 2386–2390.

Leibler, J. H., Otte, J., Roland-Holst, D., Pfeiffer, D. U., Magalhaes, R. S., Rushton, J., et al. (2009). Industrial food animal production and global health risks: Exploring the ecosystems and economics of Avian Influenza. *EcoHealth, 6*, 58–70.

Levin, S. A., Barrett, S., Aniyar, S., Baumol, W., Bliss, C., Bolin, B., et al. (1998). Resilience in natural and socioeconomic systems. *Environment and Development Economics, 3*(2), 221–262.

Levins, R. (1968). *Evolution in Changing Environments*. Princeton: Princeton University Press.

Leonard, C. (2014). *The Meat Racket: The Secret Takeover of America's Food Business*. New York: Simon & Schuster.

Leroy, E. M., Epelboin, A., Mondonge, V., Pourrut, X., Gonzalez, J. -P., Muyembe-Tamfum, J. J., et al. (2009). Human Ebola outbreak resulting from direct exposure to fruit bats in Luebo. Democratic Republic of Congo, 2007. *Vector-Borne and Zoonotic Diseases, 9*(6), 723–728. doi:10.1089/vbz.2008.0167.

Lewnard, J. A., Ndeffo Mbah, M. L., Alfaro-Murillo, J. A., Altice, F. L., Bawo, L., Nyenswah, T. G., et al. (2014). Dynamics and control of Ebola virus transmission in Montserrado, Liberia: A mathematical modelling analysis. *Lancet Infectious Diseases*. doi:10.1016/S1473-3099(14)70995-8.

Lipsitch, M., & Nowak, M.A. (1995). The evolution of virulence in sexually transmitted HIV/AIDS. Journal of Theoretical Biology, 174(4), 427–440.

Liu, Z. (2013). Dynamics of positive solutions to SIR and SEIR epidemic models with saturated incidence rates. *Nonlinear Analysis: Real World Applications, 14*, 1286–1299.

Liverani, M., Waage, J., Barnett, T., Pfeiffer, D. U., Rushton, J., Rudge, J. W., et al. (2013). Understanding and managing zoonotic risk in the new livestock industries. *Environmental Health Perspectives, 121*, 873–877. http://dx.doi.org/10.1289/ehp.1206001.

Longworth, N., Mourits, M. C., & Saatkamp, H. W. (2014). Economic analysis of HPAI control in the Netherlands II: Comparison of control strategies. *Transboundary and Emerging Diseases, 61*(3), 217–232. doi:10.1111/tbed.12034.

Luby, S. P., Gurley, E. S., & Hossain, M. J. (2009). Transmission of human infection with Nipah Virus. *Clinical Infectious Diseases, 49*, 1743–1748.

Marsh, G., Haining, J., Robinson, R., Foord, A., Yamada, M., Barr, J. A., et al. (2011). Ebola Reston virus infection in pigs: Clinical significance and transmission potential. *Journal of Infectious Disease, 204*(S3):S804–S809.

Maye, D., Dibden, J., Higgens, V., & Potter, C. (2012). Governing biosecurity in a neoliberal world: Comparative perspectives from Australia and the United Kingdom. *Environment and Planning A, 44*, 150–168.

Mayer, J. (2000). Geography, ecology and emerging infectious diseases. *Social Science & Medicine, 50*, 937–952.

McFarlane, R., Becker, N., & Field, H. (2011). Investigation of the climatic and environmental context of Hendra virus spillover events 1994–2010. *PLoS ONE, 6*(12), e28374. doi:10.1371/journal.pone.0028374.

McKenzie, F. C. (2011). *Farmer-driven innovation in agriculture: Creating opportunities for sustainability*. Dissertation, Doctor of Philosophy, School of Geosciences, Faculty of Science, University of Sydney. http://ses.library.usyd.edu.au/bitstream/2123/8924/1/McKenzie_PhDThesis_Dec2011.pdf.

Meentemeyer, R. K., Haas, S. E., & Vaclavik, T. (2012). Landscape epidemiology of emerging infectious diseases in natural and human-altered ecosystems. *Annual Review of Phytopathology, 50*, 379–402.

Mennerat, A., Nilsen, F., Ebert, D., & Skorping, A. (2010). Intensive farming: Evolutionary implications for parasites and pathogens. *Evolutionary Biology, 37*, 59e67.

Merler, S., Ajelli, M., Fumanelli, L., Gomes, M. F., Piontti, A. P., Rossi, L., et al. (2015). Spatiotemporal spread of the 2014 outbreak of Ebola virus disease in Liberia and the effectiveness of non-pharmaceutical interventions: A computational modelling analysis. *The Lancet Infectious Diseases*. pii: S1473–3099(14)71074–6. doi:10.1016/S1473-3099(14)71074-6.

Messinger, S., & Ostling, A. (2009). The consequences of spatial structure for pathogen evolution. *The American Naturalist, 174*, 441–454.

Miranda, M. E., & Miranda, N. L. (2011). Reston ebolavirus in humans and animals in the Philippines: A review. *The Journal of Infectious Diseases, 204*(3), S757–60. doi:10.1093/infdis/jir296.

Moog, F. A. (1991). The role of fodder trees in Philippine smallholder farms. In A. Speedy & P.-L. Pugliese (Eds.), *Legume trees and other fodder trees as protein sources for livestock. Proceedings of the FAO Expert consultation held at the Malaysian Agricultural Research and Development Institute (MARDI) in Kuala Lumpur, 14–18 October 1991*.

Morse, S., Mazet, J. A., Woolhouse, M., Parrish, C. R., Carroll, D., Karesh, W. B., et al. (2013). Zoonoses 3: Prediction and prevention of the next pandemic zoonosis. *The Lancet, 380*, 1956–1965.

Morvan, J. M., Nakoun, E., Deubel, V., & Colyn, M. (2000). [Forest ecosystems and Ebola virus]. *Bulletin de la Société de Pathologie Exotique, 93*(3), 172–175. [Article in French].

Murray, J. (1989). *Mathematical biology*. New York: Springer.

Murray, K. A., & Daszak, P. (2013). Human ecology in pathogenic landscapes: Two hypotheses on how land use change drives viral emergence. *Current Opinion in Virology, 3*(1), 79–83. doi:10.1016/j.coviro.2013.01.006.

Myers, K., Olsen, C. W., Setterquist, S. F., Capuano, A. W., Donham, K. J., Thacker, E. L., et al. (2006). Are swine workers in the United States at increased risk of infection with zoonotic influenza virus? *Clinical Infection and Disease, 42*, 14–20.

Mylne, A., Brad, O. J., Huang, Z., Pigott, D. M., Golding, N., Kraemer, M. U. G., et al. (2014). A comprehensive database of the geographic spread of past human Ebola outbreaks. *Scientific Data, 1*, 140042. doi:10.1038/sdata.2014.42.

Nair, G., Fagnani, F., Zampieri, S., & Evans, R. J. (2007). Feedback under data rate constraints: An overview. *Proceedings of the IEEE, 95*, 108–137.

Neumann, E. J., Kliebenstein, J. B., Johnson, C. D., Mabry, J. W., Bush, E. J., Seitzinger, A. H., et al. (2005). Assessment of the economic impact of porcine reproductive and respiratory syndrome on swine production in the United States. *Journal of the American Veterinary Medical Association, 227*, 385–392.

Nfon, C. K., Leung, A., Smith, G., Embury-Hyatt, C., Kobinger, G., & Weingartl, H. M. (2013). Immunopathogenesis of severe acute respiratory disease in Zaire ebolavirus-infected pigs. *PLoS One, 8*(4), e61904. doi:10.1371/journal.pone.0061904.

Ng, K. C., & Rivera, W. L. (2014). Antimicrobial resistance of Salmonella enterica isolates from tonsil and jejunum with lymph node tissues of slaughtered swine in Metro Manila, Philippines. *ISRN Microbiology, 2014*, 364265.

Noer, C. L., Dabelsteen, T., Bohmann, K., & Monadjem, A. (2012). Molossid bats in an African agro-ecosystem select sugarcane fields as foraging habitat. *African Zoology, 47*(1), 1–11.

Oksendal, B. (2010). *Stochastic differential equations: An introduction with applications*. New York: Springer.

Okubo, A. (1980). *Diffusion and ecological problems: Mathematical models*. Biomathematics (Vol. 10). New York: Springer.

Olival, K. J., & Hayman, D. T. (2014). Filoviruses in bats: Current knowledge and future directions. *Viruses, 6*(4), 1759–1788. doi:10.3390/v6041759.

Onwubuemeli, E. (1974). Agriculture, the theory of economic development, and the Zande Scheme. *The Journal of Modern African Studies, 12*(4), 569–587.

Otte, J., & Grace, D. (2013). Human health risks from the human-animal interface in Asia. In V. Ahuja (Ed.), *Asian livestock: Challenges, opportunities and the response. Proceedings of an International Policy Forum Held in Bangkok, Thailand, 16–17 August 2012* (pp. 121–160).

Animal Production and Health Commission for Asia and the Pacific, International Research Institute and Food and Agriculture Organization of the United Nations.

Pagiola, S., & Holden, S. (2001). Farm household intensification decisions and the environment. In R. Lee, & C. B. Barrett (Eds.), *Tradeoffs or Synergies? Agricultural Intensification, Economic Development, and the Environment.* Wallingford: CAB International.

Pan, Y., Zhang, W., Cui, L., Hua, X., Wang, M., & Zeng, Q. (2014). Reston virus in domestic pigs in China. *Archives of Virology, 159*(5), 1129–1132. doi:10.1007/s00705-012-1477-6. Epub 2012 Sep 21.

Pappalardo, M., Julia, M., Howard, M. J., Rossman, J. S., Michaelis, M., & Wass, M. N. (2016). Conserved differences in protein sequence determine the human pathogenicity of Ebolaviruses. *Scientific Reports, 6*, 23743 doi:10.1038/srep23743.

Patz, J. A., Daszak, P., Tabor, G. M., Aguirre, A. A., Pearl, M., Epstein, J., et al. (2004). Unhealthy landscapes: Policy recommendations on land use change and infectious disease emergence. *Environmental Health Perspectives, 112*(10), 1092–1098.

Pearce-Duvet, J. M. C. (2006). The origin of human pathogens: Evaluating the role of agriculture and domestic animals in the evolution of human disease. *Biological Reviews of the Cambridge Philosophical Society, 81*(3), 369–382.

Perfecto, I., & Vandermeer, J. (2010). The agroecological matrix as alternative to the land-sparing/agriculture intensification model. *Proceedings of the National Academy of Sciences, 107*(13), 5786–5791.

Peterson, A. T., Bauer, J. T., & Mills, J. N. (2004). Ecologic and geographic distribution of filovirus disease. *Emerging Infectious Diseases, 10*(1), 40–47.

Pettini, M. (2007). *Geometry and topology in Hamiltonian dynamics and statistical mechanics.* New York: Springer.

Pielou, E. (1977). *Mathematical ecology.* New York: Wiley.

Pigott, D. M., Golding, N., Mylne, A., Huang, Z., Henry, A. J., Weiss, D. J., et al. (2014). Mapping the zoonotic niche of Ebola virus disease in Africa. *eLife, 3*, e04395.

Plowright, R. K., Eby, P., Hudson, P. J., Smith, I. L., Westcott, D., Bryden, W. L., et al. (2015). Ecological dynamics of emerging bat virus spillover. *Proceedings of the Biological Sciences, 282*(1798), 20142124. doi:10.1098/rspb.2014.2124.

Protter, P. (1990). *Stochastic integration and differential equations.* New York: Springer.

Robinson, T., et al. (2014). Mapping the global distribution of livestock. *PLoS ONE, 9*(5), e96084. doi:10.1371/journal.pone.0096084.

Roden, D. (1974). Regional inequality and rebellion in the Sudan. *Geographical Review, 64*(4), 498–516.

Ruecker, N. J., Matsune, J. C., Wilkes, G., Lapen, D. R., Topp, E., Edge, T. A., et al. (2012). Molecular and phylogenetic approaches for assessing sources of Cryptosporidium contamination in water. *Water Research, 46*(16), 5135–5150.

Russell, P., & McCall, S. (1973/2013). Can succession be justified? The case of the Southern Sudan. In Wai, D. M. (Ed.), *The Southern Sudan: The problem of national integration.* Oxon: Routledge.

Saéz, A. M., Weiss, S., Nowak, K., Lapeyre, V., Zimmermann, F., Düx, A., et al. (2015). Investigating the zoonotic origin of the West African Ebola epidemic. *EMBO Molecular Medicine, 7*(1), 17–23. http://embomolmed.embopress.org/content/embomm/early/2014/12/29/emmm.201404792.full.pdf.

Sayama, Y., Demetria, C., Saito, M., Azul, R. R., Taniguchi, S., Fukushi, S., et al. (2012). A seroepidemiologic study of Reston ebolavirus in swine in the Philippines. *BMC Veterinary Research, 8*, 82. doi:10.1186/1746-6148-8-82.

Sedlock, J. L., Ingle, N. R., & Balete, D. S. (2011). Chapter 5: Enhanced sampling of bat assemblages: A field test on mount Banahaw, Luzon. *Fieldiana Life and Earth Sciences, 2*, 96–102.

Schoepp, R. J., Rossi, C. A., Khan, S. H., Goba, A., & Fair, J. N. (2014). Undiagnosed acute viral febrile illnesses, Sierra Leone. *Emerging Infectious Diseases, 20*, 1176–1182.

Sedlock, J. L., Weyandt, S. E., Cororan, L., Damerow, M., Hwa, S.-H., & Pauli, B. (2008). Bat diversity in tropical forest and agro-pastoral habitats within a protected area in the Philippines. *Acta Chiropterologica, 10*(2), 349–358. doi:http://dx.doi.org/10.3161/150811008X41492610.3161/150811008X414926.

Shafie, N. J., Sah, S. A. M., Latip, N. S. A., Azman, N. M., & Khairuddin, N. L. (2011). Diversity pattern of bats at two contrasting habitat types along Kerian River, Perak. *Tropical Life Sciences Research, 22*(2), 13–22.

Shively, G. E. (2001). Agricultural change, rural labor markets, and forest clearing: An illustrative case from the Philippines. *Land Economics, 77*(2), 268–284.

Shively, G. E., & Pagiola, S. (2001). *Poverty, agricultural development, and the environment: Evidence from a frontier region of the Philippines.* Paper presented at AAEA, Chicago, 5–8 August 2001. http://ageconsearch.umn.edu/bitstream/20532/1/sp01sh02.pdf.

Shively, G. E. & Pagiola, S. (2004). Agricultural intensification, local labor markets, and deforestation in the Philippines. Environment and Development Economics, 9 , 241–266.

Singer, P. (2005) Who pays for bird flu? Project Syndicate, 10 November 2005. http://www.project-syndicate.org/commentary/who-pays-for-bird-flu-.

Smil, V. (2002). Eating meat: Evolution, patterns, and consequences. *Population and Development Review, 28*, 599–639.

Smith, D. H., Francis, D. P., Simpson, D. I. H., & Highton, R. B. (1978). The Nzara outbreak of viral haemorrhagic fever. In S. R. Pattyn (Ed.), *Ebola Virus Haemorrhagic Fever Proceedings of an International Colloquium on Ebola Virus Infection and Other Haemorrhagic Fevers held in Antwerp, Belgium, 6–8 December, 1977.* Amsterdam: Elsevier.

Stanton, Emms & Sia (2010). *The Philippines pig farming sector: A briefing for Canadian livestock genetics suppliers.* Prepared for the Embassy of Canada in the Philippines and Office of Southeast Asia Regional Agri-Food Trade Commissioner Agriculture and Agri-Food Canada. http://www5.agr.gc.ca/resources/prod/Internet-Internet/MISB-DGSIM/ATS-SEA/PDF/5679-eng.pdf.

Stark, K. D. C. (2000). Epidemiological investigation of the influence of environmental risk factors on respiratory diseases in swine: A literature review. *Veterinary Journal, 159*, 37–56.

Stechert, C., Kolb, M., Bahadir, M., Djossa, B. A., & Fahr, J. (2014). Insecticide residues in bats along a land use-gradient dominated by cotton cultivation in northern Benin, West Africa. *Environmental Science and Pollution Research International, 21*(14), 8812–8821. doi:10.1007/s11356-014-2817-8.

Steinfeld, H., Gerber, P., Wassenaar, T., Castel, V., Rosales, M., & de Haan, C. (2006). *Livestock's Long Shadow. Environmental Issues and Options.* Food and Agriculture Organization of the United Nations, Rome.

Sun, Q., Matta, H., & Chaudhary, P. M. (2005). Kaposi's sarcoma associated herpes virus-encoded viral FLICE inhibitory protein activates transcription from HIV-1 Long Terminal Repeat via the classical NF-kappaB pathway and functionally cooperates with Tat. *Retrovirology, 2*, 9.

Taniguchi, S., Watanabe, S., Masangkay, J. S., Omatsu, T., Ikegami, T., Alviola, P., et al. (2011). Reston Ebolavirus antibodies in bats, the Philippines. *Emerging Infectious Diseases, 17*(8), 1559–1560. doi:10.3201/eid1708.101693.

Tateyama, S., Molina, H. A., Uchida, K., Yamaguchi, R., & Manuel, M.F. (2000). An epizootiological survey of necropsy cases (1993-1997) at University of the Philippines. The Journal of Veterinary Medical Science, 62(4), 439–442.

Taylor, D. J., Leach, R. W., & Bruenn, J. (2010). Filoviruses are ancient and integrated into mammalian genomes. *BMC Evolutionary Biology, 10*, 193. doi:10.1186/1471-2148-10-193.

Taylor, P. J., Monadjem, A., & Steyn, J. N. (2013). Seasonal patterns of habitat use by insectivorous bats in a subtropical African agro-ecosystem dominated by macadamia orchards. *African Journal of Ecology, 51*(4), 552–561.

Tornatore, E., Buccellato, S., & Vetro, P. (2005). Stability of a stochastic SIR system. *Physica A,* **354**(11), 111–126.

Tuckwell, H., & Williams, R. (2007). Some properties of a simple stochastic epidemic model of SIR type. *Mathematical Biosciences, 208*, 76–97.

Van Boeckel, T. P. (2013). *Intensive poultry production and highly pathogenic avian influenza H5N1 in Thailand: Statistical and process-based models*. Ph.D. Thesis, Universite Libre de Bruxelles, School of Bioengineering.

Van Boeckel, T. P., Thanapongtharm, W., Robinson, T., Biradar, C. M., Xiao, X., & Gilbert, M. (2012). Improving risk models for avian influenza: The role of intensive poultry farming and flooded land during the 2004 Thailand epidemic. *PLoS One, 7*(11), e49528. http://dx.doi.org/10.1371/journal.pone.004952810.1371/journal.pone.0049528.

Verburg, P. H., & Veldkamp, A. (2004). Projecting land use transitions at forest fringes in the Philippines at two spatial scales. *Landscape Ecology, 19*, 77–98.

Wallace, R., Wallace D., Ullmann, J. E., & Andrews H. (1999). Deindustrialization, inner-city decay, and the hierarchical diffusion of AIDS in the USA: How neoliberal and cold war policies magnified the ecological niche for emerging infections and created a national security crisis. *Environment and Planning A, 31*, 113–139.

Wallace, R., Wallace, D., & Andrews, H. (1997). AIDS, tuberculosis, violent crime and low birthweight in eight US metropolitan regions: Public policy, stochastic resonance, and the regional diffusion of inner-city markers. *Environment and Planning A, 29*, 525–555.

Wallace, R. (2002). Immune cognition and vaccine strategy: Pathogenic challenge and ecological resilience. *Open Systems and Information Dynamics, 9*, 51. doi:10.1023/A:1014282912635.

Wallace, R. (2014). Cognition and biology: Perspectives from information theory. *Cognitive Processing, 15*, 1–12.

Wallace, R. (2015). Noise and metabolic free energy in high-order biocognition. *PeerJ PrePrints, 3*, e774v1. https://peerj.com/preprints/774/.

Wallace, R., & Wallace, R. G. (2015). Blowback: New formal perspectives on agriculturally driven pathogen evolution and spread. *Epidemiology and Infection, 143*(10), 2068–2080.

Wallace, R. G. (2004). Projecting the impact of HAART on the evolution of HIV's life history. *Ecological Modelling, 176*, 227–253.

Wallace, R. G. (2009). Breeding influenza: The political virology of offshore farming. *Antipode, 41*, 916–951.

Wallace, R. G. (2014, May 8). Collateralizing farmers. *Farming Pathogens*. https://farmingpathogens.wordpress.com/2014/05/08/collateralized-farmers/.

Wallace, R. G., Bergmann, L., Hogerwerf, L., & Gilbert, M. (2010). Are influenzas in southern China byproducts of the region's globalizing historical present? In T. Giles-Vernick, S. Craddock, & J. Gunn (Eds.), *Influenza and Public Health: Learning from Past Pandemics*. London: EarthScan Press.

Wallace, R. G., Bergmann, L., Kock, R., Gilbert, M., Hogerwerf, L., Wallace, R., et al. (2015). The dawn of structural one health: A new science tracking disease emergence along circuits of capital. *Social Science & Medicine, 129*, 68–77.

Wallace, R. G., Gilbert, M., Wallace, R., Pittiglio, C., Mattioli R., & Kock R. (2014). Did Ebola emerge in West Africa by a policy-driven phase change in agroecology? *Environment and Planning A, 46*(11), 2533–2542.

Wallace, R. G., & Kock, R. A. (2012). Whose food footprint? Capitalism, agriculture and the environment. *Human Geography, 5*(1), 63–83.

Walsh, P. D., Biek, R., & Real, L. A. (2005). Wave-Like Spread of Ebola Zaire. *PLoS Biol, 3*(11), e371. doi:10.1371/journal.pbio.0030371.

Washington, M. L., & Meltzer, M. L. (2015). Effectiveness of ebola treatment units and community care centers - Liberia, September 23-October 31, 2014. *MMWR Morbidity and Mortality Weekly Report, 64*(3), 67–69.

Weingartl, H. M., Embury-Hyatt, C., Nfon, C., Leung, A., Smith, G., & Kobinger, G. (2012). Transmission of Ebola virus from pigs to non-human primates. *Scientific Reports, 2*, 811. doi:10.1038/srep00811.

Weingartl, H. M., Nfon, C., & Kobinger, G. (2013). Review of Ebola Virus infections in domestic animals. In J. A. Roth, J. A. Richt, & I. A. Morozov (Eds.), *Vaccines and diagnostics for transboundary animal diseases*, Developments in Biologicals (Vol. 135, pp. 211–218) Basel: Karger.

Wertheim, J. O., & Pond, S. L. K. (2011). Purifying selection can obscure the ancient age of viral lineages. *Molecular biology and evolution, 28*(12), 3355–3365. doi:10.1093/molbev/msr170.

WHO (2009) WHO experts consultation on Ebola Reston pathogenicity in humans. Geneva, Switzerland. 1 April 2009. Available online at http://www.who.int/csr/resources/publications/HSE_EPR_2009_2.pdf.

WHO (2015, February 4). *Ebola Situation Report*. http://apps.who.int/ebola/en/ebola-situation-report/situation-reports/ebola-situation-report-4-february-2015.

WHO Ebola Response Team (2014). Ebola virus disease in West Africa-the first 9 months of the epidemic and forward projections. *The New England Journal of Medicine, 371*, 1481–1495.

WHO/International Study Team (1978). Ebola haemorrhagic fever in Sudan, 1976. *Bulletin of the World Health Organization, 56*(2), 247–270.

Widmer, G., & Akiyoshi, D. E. (2010). Host-specific segregation of ribosomal nucleotide sequence diversity in the microsporidian *Enterocytozoon bieneusi*. *Infection, Genetics and Evolution, 10*(1), 122–128. doi:10.1016/j.meegid.2009.11.009.

Wilson, M. E., Levins, R., & Spielman, A. (Eds.) (1994). *Disease in evolution: Global changes and emergence of infectious diseases*. New York: Annals of the New York Academy of Sciences.

Wolfe, N. D., Daszak, P., Kilpatrick, A. M., & Burke, D. S. (2005). Bushmeat hunting, deforestation, and prediction of zoonoses emergence. *Emerging Infectious Diseases, 11*(12), 1822–1827.

Wolfe, N. D., Dunavan, C. P., & Diamond, J. (2007). Origins of major human infectious diseases. *Nature, 447*, 279–283.

Wood, J. L., Leach, M., Waldman, L., Macgregor, H., Fooks, A. R., Jones, K. E., et al. (2012). A framework for the study of zoonotic disease emergence and its drivers: Spillover of bat pathogens as a case study. *Philosophical Transactions of the Royal Society of London. Series B, Biological Sciences, 367*(1604), 2881–2892. doi:10.1098/rstb.2012.0228.

Xu, W., Edwards, M. R., Borek, D. M., Feagins, A. R., & Mittal, A. (2014). Ebola virus VP24 targets a unique NLS binding site on Karyopherin Alpha 5 to selectively compete with nuclear import of phosphorylated STAT1. *Cell Host Microbe, 13*, 187–200.

Yuan, J. F., Zhang, Y. J., Li, J. L., Zhang, Y. Z., Wang, L. F., & Shi, Z. L. (2012). Serological evidence of ebolavirus infection in bats, China. *Virology Journal, 9*, 236. doi:10.1186/1743-422X-9-236.

Zhang, H., Kono, H., & Kubota, S. (2014). An integrated epidemiological and economic analysis of vaccination against Highly Pathogenic Porcine Reproductive and Respiratory Syndrome (PRRS) in Thua Thien Hue Province, Vietnam. *Asian-Australasian Journal of Animal Sciences, 27*(10), 1499–1512. doi:10.5713/ajas.2014.14060.

Zhang, W., Ricketts, T. H., Kreman, C., Carney, K., & Swinton, S. M. (2007). Ecosystem services and dis-services to agriculture. *Ecological Economics, 64*, 253–260. doi:10.1016/j.ecolecon.2007.02.024.

Chapter 3
Did Neoliberalizing West Africa's Forests Produce a Vaccine-Resistant Ebola?

Robert G. Wallace, Richard Kock, Luke Bergmann, Marius Gilbert, Lenny Hogerwerf, Claudia Pittiglio, Raffaele Mattioli, and Rodrick Wallace

3.1 Is a Vaccine Enough?

Preliminary results indicate researchers have developed a successful vaccine against Ebola Makona, the *Zaire ebolavirus* variant underlying the regional outbreak in West Africa (Henao-Restrepo et al. 2015). A cluster-randomized ring vaccination trial of nearly 8000 people across Guinean location and ring size found all contacts and contacts of contacts vaccinated immediately after confirmation of a new case went uninfected. In contrast, sixteen cases emerged in those rings vaccinated 21 days after an index case.

R.G. Wallace (✉)
Institute for Global Studies, University of Minnesota, Minneapolis, MN, USA
e-mail: rwallace24@gmail.com

R. Kock
Pathology and Pathogen Biology, The Royal Veterinary College, London, UK

L. Bergmann
Department of Geography, University of Washington, Seattle, WA, USA

M. Gilbert
Biological Control and Spatial Ecology, Universite Libre de Bruxelles, Brussels, Belgium

L. Hogerwerf
Centre for Infectious Disease Control, National Institute for Public Health
and the Environment, Bilthoven, The Netherlands

C. Pittiglio • R. Mattioli
Animal Production and Health Division, Food and Agriculture Organization of the United Nations, Rome, Italy

R. Wallace
Division of Epidemiology, The New York State Psychiatric Institute, New York, NY, USA

© Springer International Publishing Switzerland 2016
R.G. Wallace, R. Wallace (eds.), *Neoliberal Ebola*,
DOI 10.1007/978-3-319-40940-5_3

55

Good news indeed, even should the vaccine prove less effective in subsequent clinical testing. Vaccines are a fundamental public health intervention when not ensnared in market failures, which are as effective a barrier to the availability of health technologies as any anti-vaxx campaign (Roush et al. 2007; Antona et al. 2013). A series of mergers and acquisitions have left only four pharmaceutical companies—GlaxoSmithKline, Sanofi-Pasteur, Merck, and Pfizer—producing vaccines for diseases other than influenza, primarily for developed markets (McNeil 2015). With little competition, many such vaccines are overpriced and effectively unavailable in the poorest countries (Pedrique et al. 2013; MacLennan and Saul 2014; Barocchi and Rappuoli 2015). The Ebola vaccine trial in West Africa was funded as a non-commercial effort by WHO, Wellcome Trust, Medecins Sans Frontieres, and the Norwegian and Canadian governments.

There is an adjunct danger in the latter success. Vaccination is based on a molecular model of disease etiology. Such thinking is necessary, of course. Viruses and immunity interact at the molecular level, even as they also do so pleiotropically, cognitively, and across multiple physiological systems (R. Wallace 2002; Van Regenmortel 2004). For a broad constituency, however, a successful vaccine implies the approach is also sufficient (Possas 2001; King 2002; Leach and Scoones 2013). An ebullient *Nature* editorial (2015), for instance, charges:

> Roll-out of the vaccine to more people will provide data to confirm its effectiveness. But by vaccinating the families, friends, health-care workers and others who come into contact with infected people, Ebola outbreaks could be stopped in their tracks the same strategy that was used to eradicate smallpox in the 1970s. This means that this vaccine can, in principle, be deployed immediately to help to end the Ebola epidemic in West Africa. As aptly conveyed by the trial's French name, *Ebola, ça suffit!* ("Ebola, that's enough!"), it is time to finish the job.

If only diseases responded to such heroic appeals to consequences alone. Many intractable pathogens, among them HIV, malaria, and tuberculosis, act decidedly unlike smallpox and other diseases that respond to the reductionist model of intervention (R. Wallace and R.G. Wallace 2004). In a world in which viruses and bacteria evolve in response to humanity's multifaceted infrastructure—agricultural, transportation, pharmaceutical, public health, scientific, and political—our epistemological and epidemiological intractabilities may be in fundamental ways one and the same.

The more socioecologically complex pathogens can evolve into population states few researchers can characterize (Gilbert and Pfeiffer 2012; R.G. Wallace et al. 2015). The financial demands placed on scientific research compound the difficulties. Models of biology and the economic doctrine under which they are produced are often tightly intertwined, down to their mathematical formalisms (Levins 2006; Schizas 2012). Many pathogens meanwhile plot their own paths, deriving solutions to interventions at one level of biocultural organization with adaptations at another (R.G. Wallace 2004). As a result, pathogen evolution routinely fails to cooperate with market expectations and scientific hypotheses alike.

3.2   Neoliberal Ebola

Ebola offers an archetypical example of such a disjunction between method and medical phenomenon.

The Makona variant appears conventional in its phenotype, if one could say so of such a dangerous pathogen, with a typical case fatality rate, incubation period, and serial interval (WHO Ebola Response Team 2014). The virus had been spilling over in the region for years. Schoepp et al. (2014) found antibodies to multiple species of Ebola in patients in Sierra Leone as far back as 5 years ago, including to the Zaire species from which the outbreak variant evolved. Phylogenetic analyses meanwhile show the species Bayesian-dated in West Africa as far back as a decade (Dudas and Rambaut 2014; Gire et al. 2014). Hoenen et al. (2015) showed the outbreak variant as initially possessing no molecular anomaly, with nucleotide substitution rates typical of Ebola outbreaks across Africa, even as Makona would phylogeographically diversify and adapt, largely by antigenic drift (Simon-Loriere et al. 2015; Carroll et al. 2015; Jun et al. 2015).

As we raise in R.G. Wallace et al. (2014), these results beg an explanation for Ebola's ecotypic shift from intermittent forest killer, taking out a village here and there, to a protopandemic infection infecting 28,000 and killing 11,000 across the region, leaving bodies in the streets of capital cities Monrovia and Conakry (WHO 2015). Even with contagion presently below replacement, the outbreak continues. Many of the thousands who survived infection suffer long-term symptomatic sequelae, including eye disease, hearing loss, arthralgia, anorexia, difficulty sleeping, and PTSD, and, as documented in one recent patient, can pass on the virus by sexual transmission (Clark et al. 2015; Qureshi et al. 2015; Reardon 2015; Christie et al. 2015).

Some commentary has noted the structural adjustment to which West Africa has been subjected the past decade included the kinds of divestment from public health infrastructure that permitted Ebola to incubate at the population level once it spilled over (Kentikelenis et al. 2015; Fallah et al. 2015). The effects, however, extend farther back in the causal chain. The shifts in land use in the Guinea Forest Region from where the Ebola epidemic first emerged were also connected to neoliberal efforts at opening the forest to global circuits of capital. It appears while Ebola did not fundamentally change, West Africa had (Gatherer 2015).

Bausch and Schwarz (2014) characterize the Forest Region as a mosaic of isolated populations of a variety of ethnic groups that hold little political power and receive little social investment. The forest's economy and ecology are also strained by thousands of refugees from civil wars in neighboring countries. The Region is subjected to the tandem trajectories of accelerating deterioration in public infrastructure and concerted efforts at private development dispossessing smallholdings and traditional foraging grounds for mining, clear-cut logging, and increasingly intensified agriculture.

The Ebola epicenter is located in the larger Guinea Savannah Zone that the World Bank describes as "one of the largest underused agricultural land reserves in the world" (Morris et al. 2009). Continental Africa hosts 60 % of the world's last farmland frontier. The Bank sees the Savannah as an opportunity best developed by market commercialization, if not solely on the agribusiness model. As the Land Matrix Observatory (2015) documents, such prospects are in the process of actualization. The Observatory lists 90 deals by which US-backed multinationals have procured hundreds of thousands of hectares for export crops, biofuels, and mining around the world, including multiple deals in Sub-Saharan Africa. The Observatory's online database shows similar land deals contracted by other world powers, including the UK, France, and China.

Under the newly democratized Guinean government, the Nevada-based and British-backed Farm Land of Guinea Limited (2011) secured 99-year leases for two parcels totaling nearly 9000 hectares outside the villages of NaeDema and Konindou in Dabola Prefecture, where a secondary Ebola epicenter developed, and 98,000 hectares outside the village of Saraya in Kouroussa Prefecture. The Ministry of Agriculture has now tasked Farm Land Inc to survey and map an additional 1.5 million hectares for third-party development. While these as of yet undeveloped acquisitions are not directly tied to Ebola, they are markers of a complex, policy-driven phase change in agroecology our group has hypothesized undergirds Ebola Makona's emergence (R.G. Wallace et al. 2014).

In an effort to connect this broader context to data accumulating about the epizoology of Ebola and the ecology of its hosts, we centered our thesis on palm oil.

Natural and semi-wild groves of different oil palm types have long served as a source of red palm oil in the Guinea Forest Region (Delarue and Cochet 2013; Madelaine et al. 2008). Forest farmers have been raising palm oil in one or another form for hundreds of years. Fallow periods allowing soils to recover, however, were reduced over the twentieth century from 20 years in the 1930s to ten by the 1970s, and still further by the 2000s, with the added effect of increasing grove density even should no new plots break ground. Concomitantly, semi-wild production has been increasingly replaced with intensive hybrids, and red oil replaced by, or mixed with, industrial and kernel oils.

Other crops are grown in the forest (Fairhead and Leach 1999; Madelaine et al. 2008). Regional shade agriculture includes coffee, cocoa, and kola. Slash-and-burn rice, maize, hibiscus, and corms of the first year, followed by peanut and cassava of the second and a fallow period, are rotated through the area. Lowland flooding supports rice. In essence, the Region has long been characterized by a move toward increased intensification without private capital in the technical sense while still remaining classifiably agroforest.

Even this passing juxtaposition has since been transformed. The Guinean Oil Palm and Rubber Company (with the French acronym SOGUIPAH) began in 1987 as a parastatal cooperative in the forest but since has grown into a state company (Delarue and Cochet 2013). SOGUIPAH is leading efforts that began in 2006 to develop plantations of intensive hybrid palm for commodity export. The company economized palm production for the market by forcibly expropriating farmland, which to this day continues to set off violent protest. During the outbreak itself, a medical team dispatched by SOGUIPAH to educate locals about Ebola and distribute chlorine was met with stones and briefly taken hostage in Bignamou, Yomou on the Liberian border (Saouromou 2015). Trust and its collapse are eminently epidemiological variables (Lomas 1998).

International aid accelerated forest industrialization. SOGUIPAH's new mill, with four times the capacity of one it previously used, was financed by the European Investment Bank (Carrere 2010). The mill's capacity ended the artisanal extraction that as late as 2010 provided local populations full employment. The subsequent increase in seasonal production has at one and the same time led to harvesting above the mill's capacity and operation below capacity off-season, leading to a conflict between the company and some of its 2000 now partially proletarianized pickers, some of whom insist on processing a portion of their own yield to cover the resulting gaps in cash flow. Pickers who insist on processing their own oil during the rainy season now risk arrest.

The new economic geography instantiates a classic case of land expropriation and enclosure, turning a tradition of shared forest commons toward expectations informal pickers working fallow land outside their family lineage obtain an owner's permission before picking palm (Madelaine 2005; Carrere 2010).

Out of the new agricultural regime an archipelago of oil palm plots has emerged in and around the Guéckédou area, the outbreak's apparent ground zero (R.G. Wallace et al. 2014). The characteristic landscape is a mosaic of villages surrounded by dense vegetation and interspersed by crop fields of oil palm and patches of open forest and regenerated young forest. The general pattern can be discerned at a finer scale as well, west of the town of Meliandou, where the index cases of the new Ebola appeared.

The landscape may embody a growing interface between humans and frugivore bats, a putative key reservoir for Ebola, including hammer-headed bats, little collared fruit bats, and Franquet's epauletted fruit bats (Pulliam et al. 2012; Olival and Hayman 2014; Plowright et al. 2015). Shafie et al. (2011) document a variety of disturbance-associated fruit bats attracted to oil palm plantations. Bats migrate to oil palm for food and shelter from the heat while the plantations' wide trails permit easy movement between roosting and foraging sites. As the forest disappears, multiple species of bat shift their foraging behavior to the food and shelter that are left.

Bushmeat hunting and butchery are one means by which subsequent spillover may take place, but agricultural cultivation may be enough of a mechanism. Anti et al. (2015) report more than a third of survey respondents in Ghana bitten by bats, scratched, or exposed to bat urine. Plowright et al. (2015) characterize bat roosting structures as conducive to indirect transmission of viruses by droplets or aerosols and warn continual exposure "may lead to a high probability of infection." Fruit bats in Bangladesh transmitted Nipah virus to human hosts by urinating on the date fruit humans cultivated (Luby et al. 2009). Even transmission by hunting may be dependent upon agriculture if by second-order effects. Leroy et al. (2009) report not long before a village outbreak large-scale hunting of Ebola-prone bats along the Lulua River in the Congo took place among the palm trees of a massive abandoned plantation bats had been visiting for half a century.

Saéz et al. (2015) have since proposed the initial Ebola spillover in Guinea occurred outside Meliandou when children, including the putative index case, caught and played with Angolan free-tailed bats in a local tree. The bats are an insectivore species also previously documented as an Ebola virus carrier. As we describe in the previous chapter, whatever the specific reservoir source, shifts in agroeconomic context still appear a primary cause (R. Wallace et al. 2016). Previous studies show the free-tailed bats also attracted to expanding cash crop production in West Africa, including sugar cane, cotton, and macadamia (Noer et al. 2012; Taylor et al. 2013; Stechert et al. 2014).

Indeed, nearly every Ebola outbreak to date appears connected to capital-driven shifts in land use, including logging, mining, and agriculture, back to the first outbreak in Nzara, Sudan in 1976, where a British-financed factory spun and wove local cotton (WHO/International Study Team 1978; Bertherat et al. 1999; Morvan et al. 2000; Groseth et al. 2007). When Sudan's civil war ended in 1972, the area rapidly repopulated and much of Nzara's local rainforests and bat ecology was reclaimed for subsistence farming, with cotton returning as the area's dominant cash crop (Roden 1974; Smith et al. 1978). As if to punctuate the point, hundreds of bats were discovered roosting in the factory itself where several workers were infected.

3.3   Structural One Health

Clearly such outbreaks are embedded beyond shifts in local ecologies brought about by the actions of specific companies in specific countries. Causality extends in space and scope. By a Structural One Health we can determine whether the world's circuits of capital as they relate to husbandry and land use, producing pronounced interconnections across the globe, are related to disease emergence (R. Wallace and R.G. Wallace 2015).

Some landscapes are enmeshed primarily within local circuits of production and exchange. Other landscapes produce traditional agricultural exports. But maps by Bergmann and Holmberg (2016) show calculations of the percentages of land (croplands, pasturelands, and forests) whose harvests are effectively consumed abroad, not only directly as agricultural goods, but also indirectly as manufactured goods and services. Further, they show how West African forests and fields are much more globalized when viewed from the perspective of the largely foreign capital investment and accumulation they directly and indirectly support, even when compared to the many overseas consumers to whose lives they contribute.

In presenting updated maps of global livestock, Robinson et al. (2014) report:

> As [agricultural] production intensifies it becomes increasingly detached from the land resource base (for example, as feeds are brought in that are grown in completely different places) and thus more difficult to predict based on spatial, agroecological variables. The effect is particularly marked for chickens and pigs, where the locations of intensive farming units often have more to do with accessibility to markets or to inputs of one sort or another, than to the agroecological characteristics of the land that can be quantified through remotely sensed variables.

If landscapes, and by extension their associated pathogens, are globalized by circuits of capital, the source of a disease may be more than merely the country in which the pathogen first appeared. As a matter of methodological completeness, we need identify which sovereign wealth funds, state-owned enterprises, governments, and private equity companies, developers, mutual funds, banks, pension funds, hedge funds, university endowments, and equity funds finance the development and deforestation leading to disease emergence in the first place (R.G. Wallace et al. 2015).

The implications are more than technical in nature, however. Such an epidemiology begs whether we might more accurately characterize such locales as New York, London, and Hong Kong, key sources of capital, as disease "hot spots" in their own right. Diseases are relational in their geographies, which are rarely confined to within the borders of a "hot zone" (Sheppard 2008; R.G. Wallace et al. 2010).

The new approach speaks to the nature of public health campaigns. The current Ebola response appears largely organized around segregating emergency operations and broader structural interventions (e.g., Osterholm et al. 2015). Emergency responses are critical, of course, but such logistics are an indirect, if perhaps in most cases unintended, means by which to avoid addressing the greater foundational contexts driving the emergence of diseases. That is, however critically unaware its practitioners, the omission serves as an ideological design feature partial to the present political and economic orders.

The philosopher Istvan Maszaros (2012) differentiates between episodic or periodic crises resolved within the established global framework and foundational crises that affect the framework itself. In the latter structural crises, unfolding in an epochal fashion through the very limits of a given order, the systemic contradictions start to accumulate in such a fashion that none can be adequately addressed. Beyond ill-defined references to "upstream" causes (e.g., Schar and Daszak 2014), we need instead explicitly acknowledge many of our emergencies, pathogens among them, arise from the very structural apparatus called upon to respond.

3.4 Forest Background Front and Center

A second false dichotomy divides pathogen and outbreak from their contextual fields. In Ebola's case, the deterministic effects of the pathogen and its evolution are treated as if divorced from the forest's ecosystemic noise—the sum of chance encounters among the various agroecological actors in the region. The reality is much more complicated, with networks of causes highly interlinked and conditional in time, space, and direction. The ostensible "background" of the forest on which Ebola and other pathogens emerge may in fact be a front and center explanation for the outbreak.

A simple stochastic differential model of exponential growth in pathogen population N can include the "noise" of stochastic ecological interactions across and within species imposed by the complexity of the forest (R.G. Wallace et al. 2014; R. Wallace and R.G. Wallace 2015),

$$dN_t = \alpha N_t dt + \sigma N_t dW_t^H \tag{3.1}$$

where $\alpha > 0$ is a characteristic rate constant for exponential growth, σ is an index of "noise" strength, and dW_t^H represents a fractional white noise process with index $0 < H < 1$. $H = 1/2$ represents ordinary white noise. The noise is defined by a covariance relationship across time and space,

$$\text{Cov}(W^H(t), W^H(s)) = 1/2(t^{2H} + s^{2H} - |t - s|^{2H}) \tag{3.2}$$

An Ito expansion produces a classic result in population growth,

$$N_t = N_0 \exp\left[\alpha t - \frac{\sigma^2}{2}t^{2H} + \sigma W_t^H\right] \tag{3.3}$$

When below a threshold, $0 < H < 1/2$ for $\alpha > 0$, the noise exponent is small enough to permit a pathogen population to explode in size. When above the threshold, for instance, $H = 1/2$ and $\sigma^2 > 2\alpha$, the noise is large enough to control an outbreak, frustrating efforts on the part of the pathogen to string together a series of susceptibles to infect above replacement.

The formalism implies under certain conditions the forest acts as its own epidemiological protection and that we risk the next deadly pandemic when we destroy that capacity. When the forest's functional noise is stripped out, the epidemiological consequences are explosive.

Control efforts are similarly impacted. Much of public health intervention, by vaccine or sanitary practices, aims at lowering an outbreak below an infection's Allee threshold, under which a population cannot reproduce enough to replace its dead (Hogerwerf et al. 2010). A pathogen, unable to find enough susceptibles to sustain itself, can be maneuvered into burning out on its own. But in this case commoditizing the forest may have lowered the region's ecosystemic threshold to such a point no emergency intervention can drive the Ebola outbreak low enough to burn out on its own. Novel spillovers suddenly express larger forces of infection. On the other end of the epicurve, a mature outbreak continues to circulate, with the potential to intermittently rebound (e.g., Barbarossa et al. 2015).

In short, neoliberalism's structural shifts are no mere background on which the emergency of Ebola takes place. The shifts are the emergency as much as the virus itself. Changes in land use brought about by policy-driven transitions in ownership and production appear fundamental contributions to explaining Ebola's area-specific emergence. Deforestation and intensive agriculture may strip out traditional agroforestry's stochastic friction, which typically keeps the virus from lining up enough transmission.

We can formalize the connections between economy and epizoology more explicitly. As presented in the previous chapter, we inductively modeled the effects of environmental stochastic noise on the resulting financial costs of an outbreak for industrial livestock, on the one hand, and agroecological production, on the other

(R. Wallace et al. 2016). We adapted Black and Scholes's (1973) approach to option pricing in finance to modeling the cost in resources needed to control epizootic outbreaks under the two models of production.

Our model shows the costs are dependent on a constant of proportionality dampening the environmental noise. If the constant is effectively zero, as occurs under agroforestry, then the cost of epidemic control grows only as the log of the policy-driven stochasticity. If the constant exceeds zero, as occurs under most industrial production, then the cost will be dominated by linear growth in the stochasticity. In short, the overall financial costs of an outbreak—including direct and opportunity costs—are dependent upon the impacts of agroeconomic policy on environmental stochasticity. The inherently explosive epizoologies of commodity agricultures—however biocontained—appear exorbitantly expensive as a first principle.

While the contention requires field testing, the Ebola outbreak in West Africa is suggestive. Bartsch et al. (2015) estimate the direct societal costs of all cases in Guinea, Liberia, and Sierra Leone through mid-December 2014 ranging from USD $82 million to $356 million.

3.5   The Political Will for a Research Way

To test these various hypotheses, we could combine remote sensing, demographic data, and trade data to spatially project the risk of another outbreak across Africa's Guinea Savannah Zone. By a number of spatial approaches, including potential surface analyses, we could project Ebola zoonotic risk across the Zone based on a number of socioecological factors, including host reservoirs, health infrastructure, human population density and mobility, shifts in land use, and globalized capital accumulation and consumption across local croplands, pasturelands, and forests, with a particular emphasis on how those factors may have evolved over time.

We could develop historical political-economic studies for the areas identified by the projection models to be at risk for novel Ebola outbreaks. Each risk area is characterized by its own place-specific social and agroeconomic trajectories. Working through local communities and supporting agencies, we could make site visits to locales already affected by outbreaks and, once the risk maps have been produced, to areas projected to be of the gravest risk. While such site visits have been previously made for Ebola, none to date has done so incorporating the broader global agroeconomics at the heart of the changes in land use behind disease spillover. Neither have such visits been made to areas of projected risk.

The question remains, however, whether in the face of current research imperatives there exists the political will to fund a project undergirded by such a set of premises. Concepts of pathogen biology can act as both a spur to and a brake upon new interventions in public health. Unwittingly or not, the new Ebola vaccine is presently applied as much as a proverbial inoculation against discussing the problems of neoliberalism's impacts upon deadly pathogens as it is a welcome addition to public health's arsenal (Degeling et al. 2015). At bottom, the two conditions are a false equivalence in practice and proposition. Blocking Ebola with a vaccine does not make the social context driving Ebola's circulation disappear. Indeed, ignoring the latter condition increases the likelihood the vaccine will fail at any number of levels, from the molecular to the socioeconomic (Van Regenmortel 2004; R. Wallace and R.G. Wallace 2004; R.G. Wallace 2008).

As Ebola and other pathogens evolve out from underneath our passing technicist responses, the agroeconomic matrix, a global specter, looms as the critical cause the health sciences are leaving largely unaddressed. That needn't be the case.

References

Anti, P., Owusu, M., Agbenyega, O., Annan, A., Badu, E. K., Nkrumah, E. E., et al. (2015). Human-bat interactions in rural West Africa. *Emerging Infectious Diseases, 21*(8), 1418–1421. doi:10.3201/eid2108.142015.

Antona, D., Lévy-Bruhl, D., Baudon, C., Freymuth, F., Lamy, M., Maine, C., et al. (2013). Measles elimination efforts and 2008–2011 outbreak, France. *Emerging Infectious Diseases, 19*, '357–364.

Barbarossa, M. V., Dénes, A., Kiss, G., Nakata, Y., Röst, G., & Vizi, Z. (2015). Transmission dynamics and final epidemic size of Ebola Virus Disease outbreaks with varying interventions. *PLoS One, 10*(7), e0131398. doi:10.1371/journal.pone.0131398.

Barocchi, M. A., & Rappuoli, R. (2015). Delivering vaccines to the people who need them most. *Philosophical Transactions of the Royal Society of London. Series B, Biological Sciences, 370*(1671). pii: 20140150. doi:10.1098/rstb.2014.0150.

Bartsch, S. M., Gorham, K., & Lee, B. Y. (2015). The cost of an Ebola case. *Pathogens and Global Health, 109*(1), 4–9.

Bausch, D., & Schwarz, L. (2014). Outbreak of Ebola virus disease in Guinea: Where ecology meets economy. *PLOS Neglected Tropical Diseases, 8*, e3056.

Bergmann, L., & Holmberg, M. (2016). Land in motion. *Annals of the American Association of Geographers, 106*, 932–956.

Bertherat, E., Renaut, A., Nabias, R., Dubreuil, G., & Georges-Courbot, M. C. (1999). Leptospirosis and Ebola virus infection in five gold-panning villages in northeastern Gabon. *The American Journal of Tropical Medicine and Hygiene, 60*(4), 610–615.

Black, F., & Scholes, M. (1973). The pricing of options and corporate liabilities. *Journal of Political Economy, 81*, 637–654.

Carrere, R. (2010). *Oil palm in Africa: Past, present and future scenarios*. Montevideo: World Rainforest Movement.

Carroll, M. W., Matthews, D. A., Hiscox, J. A., Elmore, M. J., Pollakis, G., Rambaut, A., et al. (2015). Temporal and spatial analysis of the 2014–2015 Ebola virus outbreak in West Africa. *Nature, 524*(7563), 97–101. doi:10.1038/nature14594.

Christie, A., Davies-Wayne, G. J., Cordier-Lassalle, T., Blackley, D. J., Laney, A. S., Williams, D. E., et al. (2015). Possible sexual transmission of Ebola virus - Liberia, 2015. *MMWR Morbidity and Mortality Weekly Report, 64*(17), 479–481.

Clark, D. V., Kibuuka, H., Millard, M., Wakabi, S., Lukwago, L., Taylor, A., et al. (2015). Long-term sequelae after Ebola virus disease in Bundibugyo, Uganda: A retrospective cohort study. *The Lancet Infectious Diseases, 15*(8), 905–912.

Degeling, C., Johnson, J., & Mayes, C. (2015). Impure politics and pure science: Efficacious Ebola medications are only a palliation and not a cure for structural disadvantage. *The American Journal of Bioethics, 15*, 43–45.

Delarue, J., & Cochet, H. (2013). Systemic impact evaluation: A methodology for complex agricultural development projects. The case of a contract farming project in Guinea. *European Journal of Development Research, 25*, 778–796.

Dudas, G., & Rambaut, A. (2014). Phylogenetic analysis of Guinea 2014 EBOV Ebolavirus outbreak. *PLOS Currents Outbreaks* (1st ed.). doi:10.1371/currents.outbreaks.84eefe5ce43ec9dc0bf0670f7b8b417d.

Editorial board (2015). Trial and triumph. *Nature, 524*(7563), 5.

Fairhead, J., & Leach, M. (1999). *Misreading the African landscape: Society and ecology in a forest-savanna mosaic*. Cambridge: Cambridge University Press.

Fallah, M., Skrip, L. A., d'Harcourt, E., & Galvani, A. P. (2015). Strategies to prevent future Ebola epidemics. *Lancet, 386*(9989), 131.

Farm Lands of Guinea (2011). Farm Lands of Guinea completes reverse merger and investment valuing the company at USD $45 million. PR Newswire. Available online at http://www.bloomberg.com/apps/news?pid=newsarchive&sid=a9cwc86wQ3zQ.

Gatherer, D. (2015). The unprecedented scale of the West African Ebola virus disease outbreak is due to environmental and sociological factors, not special attributes of the currently circulating strain of the virus. *Evidence-Based Medicine, 20*(1), 28. doi:10.1136/ebmed-2014-110127.

Gilbert, M., & Pfeiffer, D. U. (2012). Risk factor modelling of the spatio-temporal patterns of highly pathogenic avian influenza (HPAIV) H5N1: A review. *Spatial and Spatio-Temporal Epidemiology, 3*(3), 173–83. doi:10.1016/j.sste.2012.01.002.

Gire, S. K., Goba, A., & Andersen, K. G. (2014). Genomic surveillance elucidates Ebola virus origin and transmission during the 2014 outbreak. *Science, 345*, 1369–1372.

Groseth, A., Feldmann, H., & Strong, J. E. (2007). The ecology of Ebola virus. *Trends in Microbiology, 15*(9), 408–416.

Henao-Restrepo, A., Longini, I. M., Egger, M., Dean, N. E., Edmunds, W. J., Camacho, A., et al. (2015). Efficacy and effectiveness of an rVSV-vectored vaccine expressing Ebola surface glycoprotein: Interim results from the Guinea ring vaccination cluster-randomised trial. *Lancet.* pii: S0140-6736(15)61117-5. doi:10.1016/S0140-6736(15)61117-5.

Hoenen, T., Safronetz, D., Groseth, A., Wollenberg, K. R., Koita, O. A., Diarra, B., et al. (2015). Mutation rate and genotype variation of Ebola virus from Mali case sequences. *Science, 348*(6230), 117–119. doi:10.1126/science.aaa5646.

Hogerwerf, L., Houben, R., Hall, K., Gilbert, M., Slingenbergh, J., & Wallace, R. G. (2010). *Agroecological resilience and protopandemic influenza*. Rome: Animal Health and Production Division, Food and Agriculture Organization.

Jun, S. R., Leuze, M. R., Nookaew, I., Uberbacher, E. C., Land, M., Zhang, Q., et al. (2015, July 14). Ebolavirus comparative genomics. *FEMS Microbiology Reviews.* pii: fuv031. [Epub ahead of print]

Kentikelenis, A., King, L., McKee, M., & Stuckler, D. (2015). The international monetary fund and the ebola outbreak. *The Lancet Global Health, 3*(2), e69–70. doi:10.1016/S2214-109X(14)70377-70378.

King, N. B. (2002). Security, disease, commerce: Ideologies of postcolonial global health. *Social Studies of Science, 32*(5–6), 763–789.

Land Matrix Observatory (2015). *Global map of investments*. Available online at http://landmatrix.org/en/get-the-idea/global-map-investments/.

Leach, M., & Scoones, I. (2013). The social and political lives of zoonotic disease models: Narratives, science and policy. *Social Science & Medicine, 88*, 10e17.

Leroy, E. M., Epelboin, A., Mondonge, V., Pourrut, X., Gonzalez, J. P., Muyembe-Tamfum, J. J., et al. (2009). Human Ebola outbreak resulting from direct exposure to fruit bats in Luebo, Democratic Republic of Congo, 2007. *Vector Borne and Zoonotic Diseases, 9*(6), 723–728. doi:10.1089/vbz.2008.0167.

Levins, R. (2006). Strategies of abstraction. *Biology and Philosophy, 21*, 741–755.

Lomas, J. (1998). Social capital and health: Implications for public health and epidemiology. *Social Science & Medicine, 47*, 1181–1188.

Luby, S. P., Gurley, E. S., & Hossain, M. J. (2009). Transmission of human infection with Nipah Virus. *Clinical Infectious Diseases, 49*, 1743–1748.

MacLennan, C. A., & Saul, A. (2014). Vaccines against poverty. *Proceedings of the National Academy of Sciences of the United States of America, 111*(34), 12307–12. doi:10.1073/pnas.1400473111.

Madelaine, C. (2005). *Analyse du fonctionnement et de la dynamique de la palmeraie subspontanae en Guinae forestiare*. Cas du village de Nienh, MSc thesis. ENGREF, AgroParisTech, Montpellier.

Madelaine, C., Malezieux, E., Sibelet, N., & Manlay, R. J. (2008). Semi-wild palm groves reveal agricultural change in the forest region of Guinea. *Agroforestry Systems, 73*, 189–204.

McNeil, D. G. (2015, July 31). New meningitis strain in Africa brings call for more vaccines. *New York Times*. Available online at http://www.nytimes.com/2015/08/01/health/new-meningitis-strain-in-africa-brings-call-for-more-vaccines.html.

Maszaros, I. (2012). Structural crisis needs structural change. *Monthly Review, 63*(10), 19–32.

Morris, M. L., Binswanger-Mikhize, H. P., & Byerlee, D. (2009). *Awakening Africa's sleeping giant: Prospects for commercial agriculture in the guinea savannah zone and beyond*. Washington, DC: World Bank Publications.

Morvan, J. M., Nakoun, E., Deubel, V., & Colyn, M. (2000). [Forest ecosystems and Ebola virus]. *Bulletin de la Société de Pathologie Exotique, 93*(3), 172–175. [Article in French].

Noer, C. L., Dabelsteen, T., Bohmann, K., & Monadjem, A. (2012). Molossid bats in an African agro-ecosystem select sugarcane fields as foraging habitat. *African Zoology, 47*(1), 1–11.

Olival, K. J., & Hayman, D. T. (2014). Filoviruses in bats: Current knowledge and future directions. *Viruses, 6*(4), 1759–1788. doi:10.3390/v6041759.

Osterholm, M. T., Moore, K. A., & Gostin, L. O. (2015). Public health in the age of Ebola in West Africa. *JAMA Internal Medicine, 175*(1), 7–8. doi:10.1001/jamainternmed.2014.6235.

Pedrique, B., Strub-Wourgaft, N., Some, C., Olliaro, P., Trouiller, P., Ford, N., et al. (2013). The drug and vaccine landscape for neglected diseases (2000–11): A systematic assessment. *The Lancet Global Health, 1*(6), e371–9. doi:10.1016/S2214-109X(13)70078-0.

Plowright, R. K., Eby, P., Hudson, P. J., Smith, I. L., Westcott, D., Bryden, W. L., et al. (2015). Ecological dynamics of emerging bat virus spillover. *Proceedings of the Biological Sciences, 282*(1798), 20142124. doi:10.1098/rspb.2014.2124.

Possas, C. A. (2001). Social ecosystem health: Confronting the complexity and emergence of infectious diseases. *Cadernos de Saúde Pública, 17*(1):31–41.

Pulliam, J. R., Epstein, J. H., Dushoff, J., Rahman, S. A., Bunning, M., Jamaluddin, A. A., et al. (2012). Agricultural intensification, priming for persistence and the emergence of Nipah virus: A lethal bat-borne zoonosis. *Journal of the Royal Society, Interface, 9*(66), 89–101. doi:10.1098/rsif.2011.0223.

Qureshi, A. I., Chughtai. M., Loua, T. O., Pe Kolie, J., Camara, H. F., Ishfaq, M. F., et al. (2015). Study of Ebola Virus Disease survivors in Guinea. *Clinical Infectious Diseases*. pii: civ453. doi:10.1371/currents.outbreaks.84eefe5ce43ec9dc0bf0670f7b8b417d.

Reardon, S. (2015). Ebola's mental-health wounds linger in Africa. *Nature, 519*, 13–14.

Robinson, T. P., Wint, G. R., Conchedda, G., Van Boeckel, T. P., Ercoli, V., Palamara, E., et al. (2014). Mapping the global distribution of livestock. *PLoS One, 9*(5), e96084. http://dx.doi.org/10.1371/journal.pone.0096084.

Roden, D. (1974). Regional inequality and rebellion in the Sudan. *Geographical Review, 64*(4), 498–516.

Roush, S. W., Murphy, T. V., Vaccine-Preventable Disease Table Working Group. (2007). Historical comparisons of morbidity and mortality for vaccine-preventable diseases in the United States. *Journal of the American Medical Association, 298*, 2155–2163.

Saéz A. M., Weiss, S., Nowak, K., Lapeyre, V., Zimmermann, F., Düx, A., Kühl, H. S., et al. (2015). Investigating the zoonotic origin of the West African Ebola epidemic. *EMBO Molecular Medicine, 7*(1):17–23.

Saouromou, K. (2015). Guinae Forestiare: De nouvelles raticences a la lutte contre Ebola a Yomou. *La Express Guinee.* Available online at http://lexpressguinee.com/fichiers/videos5. php?langue=fr&idc=fr_Guinee_Forestiere_De_nouvelles_reticences_a_la_lutte_contre_.

Schar, D., & Daszak, P. (2014). Ebola economics: The case for an upstream approach to disease emergence. *EcoHealth, 11*(4), 451–452.

Schizas, D. (2012). Systems ecology reloaded: A critical assessment focusing on the relations between science and ideology. In G. P. Stamou (Ed.), *Populations, biocommunities, ecosystems: A review of controversies in ecological thinking.* Sharjah: Bentham Science Publishers.

Schoepp, R. J., Rossi, C. A., Khan, S. H., Goba, A., & Fair, J. N. (2014). Undiagnosed acute viral febrile illnesses, Sierra Leone. *Emerging Infectious Diseases, 20,* 1176–1182.

Shafie, N. J., Sah, S. A., Latip, N. S., Azman, N. M., & Khairuddin, N. L. (2011). Diversity pattern of bats at two contrasting habitat types along Kerian River, Perak, Malaysia. *Tropical Life Sciences Research, 22*(2), 13–22.

Sheppard, E. (2008). Geographic dialectics? *Environment and Planning A, 40,* 2603–2612.

Simon-Loriere, E., Faye, O., Faye, O., Koivogui, L., Magassouba, N., Keita, S., et al. (2015). Distinct lineages of Ebola virus in Guinea during the 2014 West African epidemic. *Nature, 524*(7563), 102–104. doi:10.1038/nature14612.

Smith, D. H., Francis, D. P., Simpson, D. I. H., & Highton, R. B. (1978). The Nzara outbreak of viral haemorrhagic fever. In S. R. Pattyn (Ed.), *Ebola Virus Haemorrhagic Fever Proceedings of an International Colloquium on Ebola Virus Infection and Other Haemorrhagic Fevers held in Antwerp, Belgium, 6–8 December, 1977.* Amsterdam: Elsevier.

Stechert, C., Kolb, M., Bahadir, M., Djossa, B. A., & Fahr, J. (2014). Insecticide residues in bats along a land use-gradient dominated by cotton cultivation in northern Benin, West Africa. *Environmental Science and Pollution Research International, 21*(14), 8812–8821. doi:10.1007/s11356-014-2817-8.

Taylor, P. J., Monadjem, A., & Steyn, J. N. (2013). Seasonal patterns of habitat use by insectivorous bats in a subtropical African agro-ecosystem dominated by macadamia orchards. *African Journal of Ecology, 51*(4), 552–561.

Van Regenmortel, M. H. V. (2004). Reductionism and complexity in molecular biology. *EMBO Reports, 5*(11), 1016–1020. doi:10.1038/sj.embor.7400284

Wallace, R. (2002). Immune cognition and vaccine strategy: Pathogenic challenge and ecological resilience. *Open Systems and Information Dynamics, 9,* 51. doi:10.1023/A:1014282912635.

Wallace, R., Bergmann, L., Hogerwerf, L., Kock, R., & Wallace, R. G. (2016). Ebola in the hog sector: Modeling pandemic emergence in commodity livestock. In R. G. Wallace (Ed.), *Ebola forest and farm: Neoliberal economics, environmental stochasticity, and the emergence of a deadly virus.* New York: Springer.

Wallace, R., & Wallace, R. G. (2004). Adaptive chronic infection, structured stress, and medical magic bullets: Do reductionist cures select for holistic diseases? *BioSystems, 77,* 93–108.

Wallace, R., & Wallace, R. G. (2015). Blowback: New formal perspectives on agriculturally driven pathogen evolution and spread. *Epidemiology and Infection, 143*(10), 2068–2080.

Wallace, R. G. (2004). Projecting the impact of HAART on the evolution of HIV's life history. *Ecological Modelling, 176,* 227–253.

Wallace, R. G. (2008). Book review: 'Combating the threat of pandemic influenza: drug discovery approaches'. *Quarterly Review of Biology, 83,* 327–328.

Wallace, R. G., Bergmann, L., Hogerwerf, L., & Gilbert, M. (2010). Are influenzas in southern China byproducts of the region's globalising historical present? In T. Giles-Vernick, S. Craddock, & J. Gunn (Eds.), *Influenza and public health: Learning from past pandemics.* London: EarthScan Press.

Wallace, R. G., Bergmann, L., Kock, R., Gilbert, M., Hogerwerf, L., Wallace, R., et al. (2015). The dawn of Structural One Health: A new science tracking disease emergence along circuits of capital. *Social Science & Medicine, 129,* 68–77.

Wallace, R. G., Gilbert, M., Wallace, R., Pittiglio, C., Mattioli, R., & Kock, R. (2014). Did Ebola emerge in West Africa by a policy-driven phase change in agroecology? *Environment and Planning A, 46*(11), 2533–2542.

WHO Ebola Response Team (2014). Ebola virus disease in West Africa-the first 9 months of the epidemic and forward projections. *The New England Journal of Medicine, 371*, 1481–1495.

WHO Ebola Situation Report (2015, August 12). Available online at http://apps.who.int/iris/ bitstream/10665/182071/1/ebolasitrep_12Aug2015_eng.pdf?ua=1&ua=1.

WHO/International Study Team (1978). Ebola haemorrhagic fever in Sudan, 1976. *Bulletin of the World Health Organization, 6*(2), 247–270.

Chapter 4
Introducing Pandemic Control Theory

Rodrick Wallace and Robert G. Wallace

4.1 Introduction

Conventional epidemic theory examines the spread of infectious disease as a population process driven by competition between infection growth and "removal," often with the "basic reproduction number" R_0 the central focus (e.g., Anderson and May 1991). Values > 1 determine the rate of spread in a uniform population. The Kermack–McKendrick model is typical (e.g., Bailey 1975). Let X, Y, Z be susceptible, infected, and removed populations such that $N = X + Y + Z = $ constant. Population dynamics are determined by the set of equations

$$dX/dt = -\beta XY$$

$$dY/dt = \beta XY - \gamma Y = Y(\beta X - \gamma)$$

$$dZ/dt = \gamma Y \tag{4.1}$$

Given an initial infected population $Y_0 > 0$, if $R_0 \equiv \beta X/\gamma < 1$, the infection declines to extinction. Network and stochastic elaborations abound (e.g., Bailey 1975; Anderson and May 1991; Gould and R. Wallace 1994).

A modern state, however, is not a natural population and the spread of infectious disease within it is not simply a problem in population dynamics.

Modern states are cognitive entities. Faced with dynamic patterns of threat or affordance—like spreading infection—national states or international confederations and their socioeconomic subcomponents must, can, and do

R. Wallace (✉)
Division of Epidemiology, The New York State Psychiatric Institute, New York, NY, USA
e-mail: rodrick.wallace@gmail.com

R.G. Wallace
Institute of Global Studies, University of Minnesota, Minneapolis, MN, USA

© Springer International Publishing Switzerland 2016
R.G. Wallace, R. Wallace (eds.), *Neoliberal Ebola*,
DOI 10.1007/978-3-319-40940-5_4

choose a smaller set of responses from a much larger domain of those possible to them. Such choice, in a formal manner, reduces uncertainty and implies the existence of an information source generating successive messages. That is, modern states are cognitive, and the information source or sources associated with cognitive phenomena are said to be "dual" to them (R. Wallace 2012, 2015).

How, if not by the canonical reproductive number, should disease and its cognitive context be modeled? Control theory, an interdisciplinary branch of engineering and mathematics focusing on the behavior of dynamical systems, appears more appropriate. Specifically, the Data Rate Theorem links control and information theories, tracking the systemic effects of control information constraints (Nair et al. 2007). In this case, a recent epizootic spillover or a spreading epidemic is constrained or, in the case of systemic failure, released by the adequacy or inadequacy of the control information provided by the broader social system via public health interventions or, in the longer term, by the quality of socioeconomic reform.

4.2   The Data Rate Theorem

Figure 4.1 shows a simplified schematic of a state's public health system, in the context of challenge. That is, we are assuming from the beginning that $R_0 > 1$, and that the aim is to contain the infectious disease outbreak.

The system at time t receives a multidimensional state vector X_t and produces a new vector at time $t + 1$, X_{t+1}. At time t the system is also affected by a "noise"

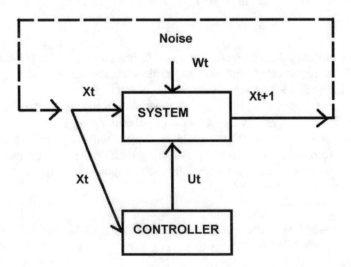

Fig. 4.1 Simplified schematic of a linear plant control system near nonequilibrium steady state. X_{t+1} is the "plant" response to the control signal U_t and the earlier state X_t. U_t is itself taken as the output of an information source. W_t is uncontrolled "noise".

vector W_t representing uncontrolled inputs, and by a "control signal" vector U_t from other cognitive entities within the state. A basic first-order "linear plant" relation near a nonequilibrium steady state can be written as

$$X_{t+1} = AX_t + BU_t + W_t \tag{4.2}$$

with feedbacks as indicated in Fig. 4.1. A and B are fixed matrices.

We suppose the system to be inherently unstable, in the sense that the matrix A can be factored by a similarity transformation into one having two diagonal submatrices A^U, A^S and two zero offdiagonal matrices such that A^U has eigenvalues ≥ 1 and A^S has eigenvalues < 1. An epidemic or pandemic contagious disease with "reproduction rate of infection" > 1 is clearly an inherently unstable system that must be brought under control by the public health institutions of a modern state via control signals U_t.

The essence of the Data Rate Theorem (Nair et al. 2007) is that the rate of control information provided by the public health system, $\mathcal{H}$, must be greater than the rate at which the unstable system generates "topological information," a relation written as

$$\mathcal{H} > \log[|\det(A^U)|] \equiv \alpha_0 \tag{4.3}$$

where det is the determinant of a matrix.

Taking the approach of Gould and R. Wallace (1994) or of R. Wallace et al. (1999), spread of infection within a modern state can be envisioned as a three-fold process of hierarchical diffusion from larger to smaller urban entities along fixed travel networks, followed by spatial diffusion from urban centers along local roads and rails according to the journey-to-work, and finally by network diffusion within smaller entities, such as families, workplaces, entertainment centers, schools, and the like. The first two stages can be fixedly characterized according to air, rail, and road networks (R. Wallace et al. 1999; Gould and R. Wallace 1994).

These constitute a fixed topology along which infection spreads. Thus the rate at which a disease generates "topological information" can be written as some composite term ρ akin to a density in a traffic model (e.g., R. Wallace 2016). Typically, for something like influenza or Ebola, ρ would be calculated as the length of a principal component from some multivariate analysis of data involving rates of deforestation, institution of plantation or factory farming, confiscation of artisanal farmlands, housing overcrowding, percent of population in poverty, rates of deindustrialization, deurbanization and violent crime (R. Wallace et al. 1999), and so on. Equation (4.3) then becomes

$$\mathcal{H}(\rho) > f(\rho)\alpha_0 \tag{4.4}$$

where α_0 now represents the inherent topology of the underlying transportation and contact networks at the scale of interest. How might we further characterize the functions $\mathcal{H}(\rho)$ and $f(\rho)$? Can we explicitly connect disease to the underlying socioeconomics?

4.3   A Black–Scholes Calculation

Take $\mathcal{H}(\rho)$ as the control information rate "cost" of public health stability at the level of environmental insult ρ. What is the mathematical form of $\mathcal{H}(\rho)$ under conditions of volatility, i.e., variability in ρ proportional to it? Let

$$d\rho_t = g(t, \rho_t)dt + b\rho_t dW_t \tag{4.5}$$

where dW_t is taken as white noise and the function $g(t, \rho)$ will "fall out" of the calculation on the assumption of certain regularities.

Let $\mathcal{H}(\rho_t, t)$ be the minimum needed incoming rate of control information under the Data Rate Theorem, and expand in ρ using the Ito chain rule (Protter 1990)

$$d\mathcal{H}_t = \left[\partial \mathcal{H}/\partial t + g(\rho_t, t)\partial \mathcal{H}/\partial \rho + \frac{1}{2}b^2 \rho_t^2 \partial^2 \mathcal{H}/\partial \rho^2 \right] dt$$
$$+ [b\rho_t \partial \mathcal{H}/\partial \rho]dW_t \tag{4.6}$$

Define a quantity L as a Legendre transform of the rate $\mathcal{H}$, by convention having the form

$$L = -\mathcal{H} + \rho \partial \mathcal{H}/\partial \rho \tag{4.7}$$

Since $\mathcal{H}$ is an information index, it is a kind of free energy in the sense of Feynman (2000) and L is a classic entropy measure.

Heuristically, replacing dX with ΔX in these expressions and applying Eq. (4.6),

$$\Delta L = \left(-\partial \mathcal{H}/\partial t - \frac{1}{2}b^2 \rho^2 \partial^2 \mathcal{H}/\partial \rho^2 \right) \Delta t \tag{4.8}$$

As in the classical Black–Scholes model (Black and Scholes 1973), the terms in g and dW_t "cancel out," and the effects of noise are subsumed into the Ito correction factor, a regularity assumption making this an exactly solvable but highly approximate model.

The conventional Black–Scholes calculation takes $\Delta L/\Delta T \propto L$. Here, at nonequilibrium steady state, we assume $\Delta L/\Delta t = \partial \mathcal{H}/\partial t = 0$, so that

$$-\frac{1}{2}b^2 \rho^2 \partial^2 \mathcal{H}/\partial \rho^2 = 0 \tag{4.9}$$

By inspection,

$$\mathcal{H} = \kappa_1 \rho + \kappa_2 \tag{4.10}$$

where the κ_i are nonnegative constants.

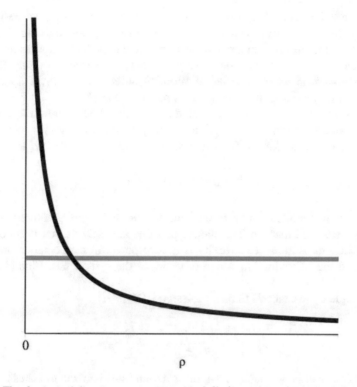

Fig. 4.2 The *horizontal line* represents the critical limit α_0. If $\kappa_2/\kappa_4 \gg \kappa_1/\kappa_3$, at some intermediate value of accumulated environmental insult ρ the relation $(\kappa_1\rho + \kappa_2)/(\kappa_3\rho + \kappa_4)$ falls below that limit, and the growth of infection becomes uncontrollable.

Taking the same level of approximation, we assume $f(\rho)$ in Eq. (4.4) can be similarly expressed as $\kappa_3\rho + \kappa_4$ so that the stability relation becomes

$$\frac{\kappa_1\rho + \kappa_2}{\kappa_3\rho + \kappa_4} > \alpha_0 \tag{4.11}$$

At low ρ the stability condition is $\kappa_2/\kappa_4 > \alpha_0$, and at high ρ it becomes $\kappa_1/\kappa_3 > \alpha_0$. If $\kappa_2/\kappa_4 \gg \kappa_1/\kappa_3$, then at some intermediate value of ρ the essential inequality may be violated, leading to uncontrolled growth of infection. See Fig. 4.2.

A second line of argument leads in a similar direction.

4.4 Information Bottleneck

A different possible approach to control system dynamics uses the information bottleneck method of Tishby et al. (1999).

The underlying conceit is that the control information needed to stabilize an inherently unstable system, which we have written as $\mathcal{H}$, can be used to define an average distortion measure in a rate distortion calculation. This requires an iterated application of the Rate Distortion Theorem (RDT) (Cover and Thomas 2006) to a control system in which a series of "control orders" $y^n = y_1, \ldots, y_n$, having probability $p(y^n)$, is sent, and the outcomes monitored as $\hat{y}^n = \hat{y}_1, \ldots, \hat{y}_n$.

The distortion measure to be associated with the RDT is now the minimum necessary control information for system stability, written as $\mathcal{H}(y^n, \hat{y}^n)$. We can thus, analogously to the standard RDT, define a new average "distortion" $\hat{\mathcal{H}}$ as

$$\hat{\mathcal{H}} \equiv \sum_{y^n} p(y^n)\mathcal{H}(y^n, \hat{y}^n) \geq 0 \qquad (4.12)$$

It is then possible to define a new, iterated Rate Distortion Function (RDF) $\mathcal{R}(\hat{\mathcal{H}})$ in a standard manner. The central point of any RDF is that it is convex in the distortion measure, and convexity is a very powerful mathematical condition determining the "direction" in which time drives the system. See Ellis (1985) for details.

For simplicity, we take $\mathcal{R}$ to be a Gaussian RDF in $\hat{\mathcal{H}}$,

$$R(\hat{\mathcal{H}}) = 1/2 \log[\sigma^2/\hat{\mathcal{H}}] \; \hat{\mathcal{H}} < \sigma^2$$
$$R(\hat{\mathcal{H}}) = 0 \; \hat{\mathcal{H}} \geq \sigma^2 \qquad (4.13)$$

Again following Feynman (2000), information must be recognized as a form of free energy and a RDF itself–as the minimum channel capacity needed to achieve a given level of average distortion–can be used to define an "entropy" as the Legendre transform

$$S = R(\hat{\mathcal{H}}) - \hat{\mathcal{H}} dR/d\hat{\mathcal{H}} \qquad (4.14)$$

Taking Onsager's nonequilibrium thermodynamics perspective (de Groot and Mazur 1984), the dynamics can, in first order, be characterized in terms of the gradients of S. We can invoke an extended analog using the stochastic differential equation (Protter 1990)

$$d\hat{\mathcal{H}}_t = \left[-\mu dS/d\hat{\mathcal{H}}_t - F(\rho)\right] dt + \beta \hat{\mathcal{H}}_t dW_t$$
$$= \left[\frac{\mu}{2\hat{\mathcal{H}}_t} - F(\rho)\right] dt + \beta \hat{\mathcal{H}}_t dW_t \qquad (4.15)$$

where dW_t is standard white noise, and $F(\rho)$ is a function of accumulated environmental insult ρ. Given an essential "topological factoring" of the control network, ρ is the only possible determinant of the rate of generation of system "topological information." β represents the magnitude of a "volatility" noise term independent of σ^2 in the definition of $\mathcal{R}$: higher $\mathcal{H}$, higher stochastic jitter.

Applying the Ito chain rule (Protter 1990) to the expectation of $\hat{\mathcal{H}}_t^2$, it becomes possible to explore the second moment stability of the system (Khasminskii 2012). A simple calculation finds that the expectation for $\hat{\mathcal{H}}^2$ cannot be a real number unless the discriminant of a quadratic equation is nonnegative, giving a necessary condition for stability as

$$F(\rho) \geq \beta \sqrt{\mu} \tag{4.16}$$

We force closure on the model by taking $F(\rho)$ as given by Eq. (4.11), so that, again,

$$\frac{\kappa_1 \rho + \kappa_2}{\kappa_3 \rho + \kappa_4} \geq \beta \sqrt{\mu} \equiv \alpha_0 \tag{4.17}$$

with similar restrictions on the constants κ_i for stability as in Fig. 4.2.

Other channel forms will have analogous stability limit conditions on ρ as a consequence of the convexity of the RDF. It is relatively easy to carry the calculation through for the "real" channel, having $\mathcal{R}(\hat{\mathcal{H}}) = \sigma^2/\hat{\mathcal{H}}$.

4.5 Turbulence

A third line of argument, adapted from the air traffic control model of Hu et al. (2001), involves geodesic streamline flow in a topological quotient space for infection spread along travel networks. Given the inherent instability of pandemic propagation, "acceptable" dynamic trajectories can be envisioned as constrained to geodesic streamlines in a large socioeconomic "biospace" $\mathcal{M}$ that have been factored according to a maximum infection rate r per unit population.

The equivalence class of such streamlines is written as $W(r)$, and the topological quotient space as $\mathcal{M}/W(r)$, which may be very large indeed. Deviations from geodesics in this quotient space represent socioeconomic infection trajectories having rates per unit population greater than r. Let K_t be an appropriate index of the degree of deviation from such an "acceptable" geodesic trajectory at time t. We can express the dynamics of K_t, in first order, by the stochastic differential equation

$$dK_t \approx aK_t dt + \sigma K_t dW_t \tag{4.18}$$

where $a, \sigma \geq 0$ and dW_t represents white noise. Applying the Ito chain rule to $\log[K_t]$, we obtain

$$d \log[K_t] \approx (a - \sigma^2/2)dt + \sigma dW_t \tag{4.19}$$

so that, if $2a \equiv \alpha_0 < \sigma^2$, the expectation $E[K]_t \to 0$. Our assertion is that the "noise" σ^2 is again $\mathcal{H}(\rho)/f(\rho)$ from Eq. (4.4), since ρ is the only possible index for the generation of topological information by socioeconomic or political instability, and $\mathcal{H}$ is the control signal. *Everything else has been factored out through the construction of geodesics in* $\mathcal{M}/W(r)$. We again invoke the closure relation of Eq. (4.11) and the implications of Fig. 4.2.

The "characteristic area" method of Gould and R. Wallace (1994) provides a case history involving the propagation of AIDS through the commuting field near New York City.

A slightly more sophisticated public health control model might use the product of r, the infection rate per unit population, and its time rate of growth, dr/dt, i.e., an index $\hat{r} \propto dr^2/dt$, leading to an equivalence class $W(\hat{r})$ and a quotient topology on $\mathcal{M}/W(\hat{r})$.

4.6 Cognitive Symmetry Breaking

A deeper approach to the dynamics of infection regulation is via the "cognitive paradigm" of Atlan and Cohen (1998), who recognized that the immune response is not simply an automatic reflex, but involves active choice of a particular response to insult from a larger repertoire of those possible to it. As described, such choice reduces uncertainty in a formal manner, and implies the existence of an information source (R. Wallace 2012, 2015).

Given an information source associated with an inherently unstable infection/public health system—characterized as "dual" to it—an equivalence class algebra can be constructed by choosing different system origin states a_0 and defining the equivalence of two subsequent states at times $m, n > 0$, written as a_m, a_n, by the existence of high-probability meaningful paths connecting them to the same origin point. Disjoint partition by equivalence class, essentially similar to orbit equivalence classes in dynamical systems, defines a symmetry groupoid associated with the cognitive process. Groupoids represent generalizations of the group concept in which there is not necessarily a product defined for each possible element pair (Weinstein 1996). The simplest example would be a disjoint union of groups.

The equivalence classes define a set of cognitive dual information sources available to the inherently unstable public health/infectious disease system, creating a large groupoid, with each orbit corresponding to a transitive groupoid whose disjoint union is the full groupoid. Each subgroupoid is associated with its own dual information source, and larger groupoids will have richer dual information sources than smaller.

Let X_{G_i} be the public health control system's dual information source associated with the groupoid element G_i, and let Y be the information source associated with incoming environmental stress, in a large sense. See R. Wallace (2012, 2015) for details of how environmental regularities imply the existence of an environmental information source, in the broad sense of the concept as described in R. Wallace (2012).

As is standard in statistical mechanics models of phase transitions, we construct a Morse Function (Pettini 2007) as follows.

Let $H(X_{G_i}, Y) \equiv H_{G_i}$ be the joint uncertainty of the two information sources. Define a Boltzmann-like pseudoprobability as

$$P[H_{G_i}] = \frac{\exp[-H_{G_i}/\kappa\mathcal{H}]}{\sum_j \exp[-H_{G_j}/\kappa\mathcal{H}]} \tag{4.20}$$

where κ is an appropriate constant depending on the particular system and its linkages to control signals, and the sum is over the different possible cognitive modes of the full system.

A "free energy" Morse Function F can be defined as

$$\exp[-F/\kappa\mathcal{H}] \equiv \sum_j \exp[-H_{G_j}/\kappa\mathcal{H}] \tag{4.21}$$

Given the inherent groupoid structure, it is possible to apply an extension of Landau's picture of phase transition (Pettini 2007). In Landau's formulation of spontaneous symmetry breaking, phase transitions driven by temperature changes occur as alteration of system symmetry, with higher energies at higher temperatures being more symmetric. The shift between symmetries is highly punctuated in the temperature index, here the minimum necessary control information rate $\mathcal{H}$ under the Data Rate Theorem for unstable control systems. Typically, such arguments involve only a very limited number of possible phases. In this context such phases would divide—as in a physical system either water or ice—into the infection contained or not contained.

Decline in the richness of control information $\mathcal{H}$, or in the ability of that information to influence the system, characterized by κ, can lead to punctuated decline in the complexity of cognitive process possible within the public health control system, driving it into a ground state collapse in which infection proliferates beyond the acceptable rate r (or the expanded index $\hat{r} \propto dr^2/dt$).

The essential feature is the integrated environmental insult ρ. Most of the topology of the inherently unstable public health/infection system has been "factored out" via the construction of geodesics in a topological quotient space $\mathcal{M}/W(r)$ or $\mathcal{M}/W(\hat{r})$—inherently incorporating the topology of the underlying travel networks—so that ρ remains the only possible index of the rate of topological information generation for the DRT. Thus, in Eqs. (4.20) and (4.21), $\mathcal{H}$ is again replaced by the ratio $\mathcal{H}(\rho)/f(\rho)$, where f is a dimensionless monotonic increasing positive function.

For a fixed $\mathcal{H}$, increasing ρ is then equivalent to lowering the "temperature," and the system passes from high symmetry "free flow" to different forms of "crystalline" structure—broken symmetry representing onset of significant infection spread.

Again, making an exactly solvable Black–Scholes approximation, the "cost" of the control information $\mathcal{H}$ can, in first order, be expressed in terms of a linear function of ρ and again, in first order, taking $f(\rho) \approx \kappa_3\rho + \kappa_4 > 0$, we obtain an effective "temperature" as

$$\kappa \mathcal{H}(\rho)/f(\rho) \approx \frac{\kappa_1\rho + \kappa_2}{\kappa_3\rho + \kappa_4} \tag{4.22}$$

If $\kappa_2/\kappa_4 \gg \kappa_1/\kappa_3$, accumulated environmental insult will quickly bring the effective "temperature" below some critical value, triggering collapse into a dysfunctional ground state of spreading infection.

4.7 Discussion and Conclusions

Modern states are cognitive entities incorporating lower-level control systems that interact across a variety of scales and levels of organization. The spread of infection within such an entity is not simply a matter of "population dynamics." Rather, the intertwining of social, economic, and political historical trajectories creates "riverbanks" that constrain the constant flow of contextual public health decision-making that is necessary for epidemic control and containment (R. Wallace and Fullilove 2008).

Factoring out the underlying "topology" of those riverbanks, including network travel paths, may permit identification of a composite environmental burden index ρ that drives pandemic outbreaks. For swine and avian influenza worldwide, it appears that confined feedlot animal facilities—factory farms—constitute the central mechanisms generating pandemic threat (R. Wallace and R.G. Wallace 2015). For Ebola in West Africa, the spread of plantation farming for palm oil and other cash crops appears central (R.G. Wallace et al. 2014). The current Zika virus outbreak across South and Central America is linked to cutbacks in mosquito control expenditures in the context of global climate change, continuing deforestation, and latifundia-led land consolidation (Weaver and Reisen 2010; Chaves 2013).

All of these dynamics are matters of public policy and economic practice that are quintessentially cognitive, in the particular context of "structural adjustments" and unconstrained neoliberal agroecologies. Travel patterns underlying disease spread are themselves strongly defined by the circuits of capital associated with those agroecologies and their parent political socioeconomies.

Figure 4.2 demarcating the critical value at which environmental perturbation leads to collapse in disease control suggests that apparently "small" increments in socioecological expropriation can rapidly erode the ability of public health systems to contain deadly pathogens. It follows that a key step toward pandemic prevention is to reverse capital-led extraction and structural adjustment, a dual program in neoliberal institutional cognition. By logging, mining, and intensive agriculture deep into the forest, the former depresses long-term ecosystemic controls, permitting previously marginalized pathogens regional and potentially pandemic escape. The latter program truncates public and animal health, softening herd immunity for newly emergent or suddenly resurgent diseases.

References

Anderson, R., & May, R. (1991). *Infectious diseases of humans: Dynamics and control*. New York: Oxford University Press.

Atlan, H., & Cohen, I. (1998). Immune information, self-organization and meaning. *International Immunology, 10*:711–717.

Bailey, N. T. J. (1975). *The mathematical theory of infectious diseases and its applications*. New York: Hafner.

Black, F., & Scholes, M. (1973). The pricing of options and corporate liabilities. *Journal of Political Economy, 81*:637–654.

Chaves, L. (2013). The dynamics of Latifundia formation. *PlosONE, 8*(12), e8263.

Cover, T., & Thomas, J. (2006). *Elements of information theory*. New York: Wiley.

de Groot, S., & Mazur, P. (1984). *Non-Equilibrium thermodynamics*. New York: Dover.

Ellis, R. (1985). *Entropy, large deviations and statistical mechanics*. New York: Springer.

Feynman, R. (2000). *Lectures on computation*. Boulder, CO: Westview.

Gould, P., & Wallace, R. (1994). Spatial structures and scientific paradoxes in the AIDS pandemic. *Geografiska Annaler, 76B*, 105–116.

Hu, J., Prandini, M., Johansson, K., & Sastry, S. (2001). Hybrid geodesics as optimal solutions to the collision-free motion planning problem. In M. Di Benedetto & A. Sangiovanni-Vincetelli (Eds.), *Hybrid systems: Computation and control, HSCC 2001*. Lecture Notes in Computer Science, Springer-Verlag, Berlin/Heidelberg: Springer (Vol. 2034, pp. 305–318).

Khasminskii, R. (2012). *Stochastic stability of differential equations*. New York: Springer.

Nair, G., Fagnani, F., Zampieri, S., & Evans, R. (2007). Feedback control under data rate constraints: An overview. *Proceedings of the IEEE, 95*, 108–114.

Pettini, M. (2007). *Geometry and Topology in Hamiltonian Dynamics*. New York: Springer.

Protter, P. (1990). *Stochastic Integration and Differential Equations*. New York: Springer.

Tishby, N., Pereira, F., & Bialek, W. (1999). The information bottleneck method. *37th Allerton Conference on Communication, Control and Computing*, pp. 368–377.

Wallace, R. (2012). Consciousness, crosstalk and the mereological fallacy: An evolutionary perspective. *Physics of Life Reviews, 9*, 426–453.

Wallace, R. (2015). *An information approach to mitochondrial dysfunction: Extending Swerdlow's hypothesis*. Singapore: World Scientific.

Wallace, R. (2016). Canonical instabilities of autonomous vehicle systems. https://peerj.com/preprints/1714/.

Wallace, R., & Fullilove, M. (2008). *Collective consciousness and its discontents*. New York: Springer.

Wallace, R., Ullmann, J., Wallace, D., & Andrews, H. (1999). Deindustrialization, inner city decay and the hierarchical diffusion of AIDS in the US: How neoliberal and cold war policies magnified the ecological niche for emerging infections and created a national security crisis. *Environment and Planning A, 31*, 113–139.

Wallace, R., & Wallace, R. G. (2015). Blowback: New formal perspectives on agriculturally driven pathogen evolution and spread. *Epidemiology and Infection, 143*(Special Issue)(10), 2068–2080.

Wallace, R. G., Gilbert, M., Wallace, R., Pittiglio, C., Mattioli, R., & Kock, R. (2014). Did Ebola emerge in West Africa by a policy-driven phase change in agroecology? *Environment and Planning A, 46*, 2533–2542.

Weaver, S., & Reisen, W. (2010). Present and future arboviral threats. *Antiviral Research, 85*, 328–345.

Weinstein, A. (1996). Groupoids: Unifying internal and external symmetry. *Notices of the American Mathematical Association, 43*, 744–752.

Chapter 5
The Social Amplification of Pandemics and Other Disasters

Rodrick Wallace and Robert G. Wallace

5.1 Introduction

A disaster is a relatively rapid event causing large-scale population-level, as opposed to individual, stress. Conflagration, economic collapse, widespread organized violence, extreme weather or volcanic conditions, plague outbreaks, and so on, all provide cogent examples. While our particular focus here is upon pandemics, other disastrous incidents can provide examples of the mechanisms by which social disruption triggers avalanches of additional morbidity and mortality.

With regard to pandemics in particular, a growing public and animal health literature suggests current neoliberal patterns of agroeconomic exploitation and expropriation will in all likelihood set off a deadly pandemic of an RNA virus or other pathogen (Leibler et al. 2009; R.G. Wallace 2009; Mennerat et al. 2010; Drew 2011; Ercsey-Ravasz et al. 2012; Jones et al. 2013; Liverani et al. 2013; Engering et al. (2013); FAO et al. 2013; Khan et al. 2013; Kentikelenis et al. 2014; R.G. Wallace et al. 2015).

Ecosystems in which "wild" virus circulate and are controlled by the rough-and-tumble of environmental stochasticity are being suddenly streamlined by way of deforestation and plantation monoculture (e.g., R.G. Wallace et al. 2014). Pathogen spillovers that once died out relatively quietly are now finding the chain of susceptibles necessary to turn into outbreaks of great extent, duration, and momentum. These are entrained into large-scale confined feedlot animal facilities—pejoratively, factory farms—where wild types are domesticated into livestock and

R. Wallace (✉)
Division of Epidemiology, The New York State Psychiatric Institute, New York, NY, USA
e-mail: rodrick.wallace@gmail.com

R.G. Wallace
Institute for Global Studies, University of Minnesota, Minneapolis, MN, USA

© Springer International Publishing Switzerland 2016
R.G. Wallace, R. Wallace (eds.), *Neoliberal Ebola*,
DOI 10.1007/978-3-319-40940-5_5

proximate human populations (R. Wallace et al. 2016). One or more of these pathogens are expected to escape local extirpation to become the next "1918" pandemic, with global spread and high rates of incapacitation and mortality.

What appears less well understood is that such a pandemic, devastating as it may be, can represent only the beginning of a greater and longer-term policy-driven disaster—defined as a man-made event causing large-scale population stress. The concept of catastrophic path dependency is old hat in other domains. By a convergence of ideological commitments and material circumstance, the 9/11 attack on New York City was treated as a rationale for the subsequent occupation of Iraq, leading to enduring low level conflict and the religious fanaticism now afflicting the region. Similar excess followed the 1980s Soviet invasion of Afghanistan a generation earlier, ironically counting among its sequalae the mujahideen's 9/11 blowback.

One of the earliest published analyses of such a cascade is a landmark 1962 *New England Journal of Medicine* issue on the health effects of thermonuclear war. Contributors examined continuing loss-of-life in the post-attack period following a 10-weapon, 56 megaton strike on the Boston area. The initial burst and subsequent firestorms were estimated as killing 2.2 million. But, as Ervin et al. (1962) pointed out:

> The longer-term survival of human populations after this ecologic upheaval would be precarious. Even assuming an intact social structure and the maintenance of a functioning workforce, agriculture, particularly domestic animals, would be all but destroyed. Before malnutrition became a major medical concern, however, the threat of epidemic infectious disease would be raised by the fact that bacteria, fungi, viruses and insects would survive the effects of radiation. The ultimate size of these populations in the absence of challenge by their natural enemies is difficult to estimate.

Human survivors would likely face such infectious diseases in the context of immune dysfunction consequent on psychosocial stress that becomes synergistic with persistent radiation and combustion product poisoning, then followed by famine.

Here, moving past scenarios, we will first present an actual large-scale human ecological collapse in 1970s New York City. In an effort to break minority voting blocs, New York's municipal government withdrew essential municipal services from overcrowded, low income neighborhoods of color. The effects, extending across whole neighborhoods, ignited a rapid large-scale process of contagious fire, building abandonment, and forced population displacement that had surprisingly long-term impacts on patterns of public health and public order extending up to the national scale (D. Wallace 2001; D. Wallace and R. Wallace 1998, 2011; Gould and R. Wallace 1994; R. Wallace et al. 1999).

The degree of damage in New York City, while not approaching NEJM's nuclear holocaust scenario, approximated the aftermath of massive, concentrated conventional bombing.

It is even possible to make a back-of-the-envelope calculation of the social amplification factor associated with such an anthropogenic disaster. We will apply the insights from this example to pandemics of agroecological origins, using a standard epidemic model and setting the impact of 1918 influenza as a boundary condition.

The implications are broadly applicable: contrasting with the more familiar "social amplification of risk" perspective of Kasperson et al. (1988), we focus on the cascading socioecological morbidities and mortalities triggered by a seminal mass-fatal event.

5.2   The New York City Fire Epidemic

Television coverage of the 1977 Baseball World Series in New York City, showing local tenements on fire around Yankee Stadium, brought the phrase "the Bronx is burning" into public consciousness. In reality, crowded minority neighborhoods of New York had suffered high levels of fire and building abandonment since the late 1960s.

Figure 5.1 shows a citywide index of building fire damage based on official fire department statistics (D. Wallace and R. Wallace 1998, 2011; R. Wallace and D. Wallace 1977, 1990). Of particular note is the sharp rise between 1967 and 1968, followed by a plateau and decline through 1972. This was the result of the opening of some twenty new fire companies as "second sections" of existing fire units in high fire incidence areas that was forced under an arbitration agreement awarded to the city's fire service unions, the Uniformed Fire Officers and Uniformed Firefighters Associations. The new fire companies served much as a boost to an immune system facing an infectious disease, allowing more immediate suppression of visible fire damage and permitting landlords to rebuild rather than abandon their damaged buildings. For the details of contagious building fire and abandonment, and their relation to fire service levels, see D. Wallace (2001), D. Wallace and R. Wallace (1998, 2011) or R. Wallace and D. Wallace (1977, 1990).

In essence, increased fire extinguishment services protected significant levels of badly overcrowded housing in poor neighborhoods. Cuts in fire service, justified by simplistic mathematical models developed by the Rand Corporation at the request of New York City (D. Wallace and R. Wallace 2011; R. Wallace and D. Wallace 1977), reduced epidemiologic threshold below existing rates of overcrowded housing, setting the stage for contagious fire, and abandonment to proceed to completion. Indeed, in the Bronx, the 1970 synergism between poverty and housing overcrowding closely predicted subsequent burnout (R. Wallace 1990). Figure 5.2 shows, for the Bronx, the integral under the curve of Fig. 5.1, 1970–1980, as measured across Health Areas, which are the small aggregates of Census Tracts by which health statistics are reported. Some sections of the borough—one of the largest conurbations in the Western world—lost more than 80 % of their occupied housing units in that time.

Other areas of the city—Central Harlem in Manhattan, Bushwick, Brownsville, East New York in Brooklyn, Jamaica and South Jamaica in Queens—suffered

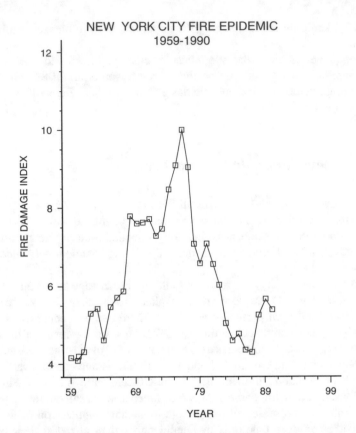

Fig. 5.1 The epidemic curve for contagious fire and building abandonment in New York City, 1959–1990. Note the stabilization between 1968 and 1972 caused by significant increase in fire extinguishment service, and the subsequent sharp rise caused by the closing of some fifty firefighting units and the destaffing of those that remained. The decline after the peak represents the wholesale loss of housing units, causing forced population shifts that drove those with resources out of the city, leaving their evacuated housing units to displaced refugees. This brought the number of badly overcrowded housing units below epidemiologic threshold.

similar devastation as a result of fire service reductions targeted against them (D. Wallace and R. Wallace 1998, 2011). Citywide, some 1.2 million persons were directly and indirectly displaced, with some 600,000 remaining behind in the devastated zones. Photographs from this time show large areas that resembled Berlin after World War II, an unprecedented circumstance for a Western nation outside of wartime conditions.

In reality, historical analysis shows this devastation to be the deliberate outcome of policy aimed at dispersing minority political power in the city, and stalling the advance of the Civil Rights Movement from the American South into the heavily segregated cities of the North (D. Wallace and R. Wallace 1998). The mathematical models of the Rand Corporation used to justify these policies, while ludicrous on

PERCENT HOUSING LOST 1970–80

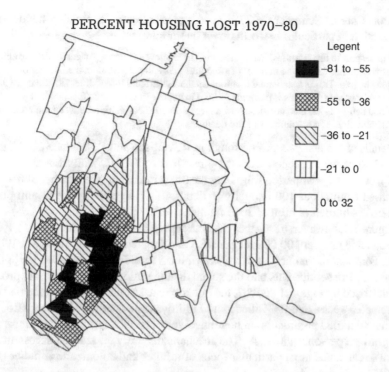

Legent

■ −81 to −55

▨ −55 to −36

▧ −36 to −21

▥ −21 to 0

☐ 0 to 32

Fig. 5.2 Percent occupied housing lost in Bronx Heath Arcas, 1970–1980. Large regions of a conurbation containing 1.4 million people lost over half their housing units. Subsequent massive population displacement dismembered community structures among both those displaced and those forced to leave the overwhelmed receiving neighborhoods (D. Wallace 2001).

their face from a fire service perspective (R. Wallace and D. Wallace 1977; D. Wallace and R. Wallace 2011), were sufficient to prevent accusations of "arbitrary and capricious behavior" to proceed under US law, and smoothed the way for de-facto ethnic cleansing carried out under color of law. A complaint to the US Attorney for the Southern District in the waning days of the Carter Administration elicited the comment by an FBI agent assigned to these matters by the Civil Rights Division of the Justice Department that the city was clearly "trying to break up voting blocs" using fire service reductions as the principal tool, perhaps the most definitive professional opinion on the matter to date.

Although the public health consequences of this event have been a matter of some study (R. Wallace and D. Wallace 1990; D. Wallace and R. Wallace 1998, 2011; R. Wallace 1990; D. Wallace 1994, 2001), for the purposes of a larger analysis we wish to compare direct and indirect loss of life, and will infer the general from the particular of this case history.

Direct loss of life to the New York City fire epidemic must be indirectly estimated R. Wallace and D. Wallace (1977), since fire death statistics were systematically falsified by the fire department. In September 1977, then-Assemblyman (now US

Senator) Charles Schumer (1977) issued a press release denouncing the deception. One section in particular is worth direct quotation,

> With respect to fire fatalities, the figures after 1972 show an almost perfect (0.99) correlation between an increase in the number of structural fires and a *decrease* in the number of people dying in fires. This is a statistical relationship that is virtually impossible: i.e., the more fires, the fewer people die. This is particularly hard to believe in view of the fact that in the 10-year period before 1972, the statistics show a strong *positive* correlation between the number of structural fires and the number of fire deaths.

Details of the analysis can be found in R. Wallace and D. Wallace (1977, p. 30).

As best as we can reconstruct the direct loss of life (R. Wallace and D. Wallace 1977), over a 10-year period following the fire service cuts, the number of fire deaths increased from about 100 annually to about 400, suggesting a maximum possible total excess burden of around 3000 fatalities.

Figure 5.3, however, raises the stakes. Adapted from Monkkonen (2001), it shows the homicide rate per 100,000 for the USA and for New York City between 1900 and 1998. For most of that period, NYC lags the US rate. During and after the "planned shrinkage" fire service cuts and the social disintegration that followed the processes characterized by Figs. 5.1 and 5.2, the rate rose catastrophically, and remained raised for some 20 years. See R. Wallace and Fullilove (2014) or R. Wallace (2015, Ch 8) for more detailed arguments on how massive social disintegration can cause raised rates of interpersonal violence. The mechanisms are well studied across much of social science, and are a particular focus of refugee and colonization studies (Fanon 1966).

Of central interest is the two-peak form of Fig. 5.3. The first peak seems associated with the direct consequences of urban burnout. The second peak, almost a decade later, represents the "crack wars" consequent on the persisting destabilization of the illegal drug trade (R. Wallace and Fullilove 2014).

The direct excess loss of life is represented by an increase from a base of about 500 annual homicides to about 2000, over a 20-year period, suggesting a primary burden of some 30,000 excess deaths. Assuming another 30,000 were badly injured and displaced onto truncated life course trajectories via substance abuse, behavioral reaction pathologies, and similar mechanisms, there were perhaps as many as 60,000 premature mortalities consequent on the violence epidemic that followed the events of Figs. 5.1 and 5.2.

More deaths, however, can be associated with New York City's ethnic cleansing efforts. As one public official put the matter, "Planned shrinkage shotgunned AIDS over the Bronx." It also shotgunned AIDS over Manhattan, Brooklyn, and Queens, and perhaps, since New York City sits atop the US urban hierarchy, nationally as well (Gould and R. Wallace 1994; R. Wallace et al. 1999). The total burden in premature mortalities resulting from the policies causing Figs. 5.1 and 5.2 may, over a 30-year period, exceed 100,000, in comparison with the 3000 direct fire deaths. Certainly an order of magnitude increase due to social amplification seems likely, for this case history.

D. Wallace (2001) explored in some detail the subsequent New York City tuberculosis epidemic:

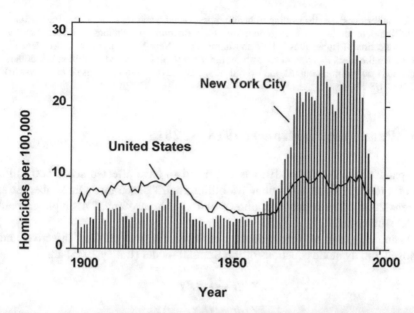

Fig. 5.3 The social amplification of disaster. Figure adapted from Monkkonen (2001), showing the homicide rate per 100,000 for the USA and NYC, 1900–1998. After the devastation following "planned shrinkage" fire service cuts, the number of NYC homicides rose from about 500–2000 per year for 20 years, an excess of some 30,000 direct fatalities. Supposing another 30,000 were badly injured but not directly killed, and put on a deteriorating life course trajectory, there may have been as many as 60,000 premature fatalities over a 30-year period following the fire service reductions. Note the two separate NYC peaks. The first is coincident with the fire epidemic. The second, *larger peak*, represents the "crack wars" consequent on destabilization of illegal drug marketing pattern and practice by the forced migration associated with large-scale urban desertification. Other premature mortalities may have followed the shotgunning of AIDS and contagious behavioral pathologies across the city, the metropolitan region, and down the US urban hierarchy.

> At the height of the fire epidemic…the first signs that control over tuberculosis had weakened in New York City appeared: large increases in incidence 2 years in a row, consecutive large increases unprecedented since World War II…The burnout of the old poor neighborhoods was a disaster in the true sense of the word, a horror like war or an earthquake. The psychological impact resembled that of a disaster, including the problem behaviors such as substance abuse, violence, and promiscuity. This constellation of problem behaviors constituted an escalating behavioral code, a behavioral language among the youth to differentiate and establish identity in the disorganized, traumatized areas. Both alcohol and illegal drugs affect the immune system and have historically been associated with high incidence of tuberculosis. The drugs and promiscuity led to new tuberculosis cases directly and indirectly through transmission of the AIDS virus, HIV. Infection with TB facilitates infection with HIV; infection with HIV greatly facilitates infection with TB and rapid progression to active disease…

Freudenberg et al. (2006)—writing some time after this—concluded that New York City's "planned shrinkage" program indeed had massive health impacts:

Cuts in services; the dismantling of health, public safety, and social service infrastructures and the deterioration of living conditions for vulnerable populations contributed to the amplification of [tuberculosis, HIV and homicide in New York City] over 2 decades. We estimate the costs incurred in controlling these epidemics exceeded 50 billion US dollars (in 2004 dollars); in contrast, the overall budgetary saving during the fiscal crisis was 10 billion US dollars.

5.3 Pandemic Penetrance: 1918 vs. 2015

The pandemic influenza of 1918 is estimated to have affected some 500 million out of a total population of about 1.5 billion, a 1/3 penetrance (Taubenberger and Morens 2006). The benchmark allows us to calibrate Kendall's simple epidemic model with removal.

Assuming a total of N individuals, classified as X susceptible, Y infective, and Z removed, the dynamic equations of the Kendall model (Bailey 1975) are

$$dX/dt = -\beta XY$$
$$dY/dt = (\beta X - \gamma)Y$$
$$dZ/dt = \gamma Y$$
$$N = X + Y + Z \tag{5.1}$$

β is the rate of infection and γ the removal rate. Letting $\rho = \gamma/\beta$, no epidemic can spread if the removal rate is greater than the infection rate, i.e., $X(0) < \rho$, where $X(0)$ is the susceptible population at time $t = 0$, and N the total population.

Following Kendall, (Bailey 1975, Eq. 6.22) shows that, if I is the proportion of the total number of susceptibles that finally contract the disease, assuming a small initial number of infectives and a large N, then

$$N/\rho \equiv s = -\log[1 - I]/I \tag{5.2}$$

This has the solution

$$I = 1 + \frac{W[-s\exp(-s)]}{s} \tag{5.3}$$

where $s = N/\rho$ and W is the Lambert W-function, which solves the relation $x = W(x)\exp[W(x)]$. Equation (5.3), however, also represents the size of the "giant component" in a random-network percolation model (e.g., Parshani et al. 2010; Gandolfi 2013). See Fig. 5.4, which shows I as a function of the ratio N/ρ.

A simple calculation can be based on the 1918 observations as a boundary condition. In 1918, some 1/3 of the total population of 1.5 billion became infected, suggesting a critical population ρ of about 1.233 billion, under 1918 travel conditions. By 2015, the total population has reached about 7 billion, suggesting

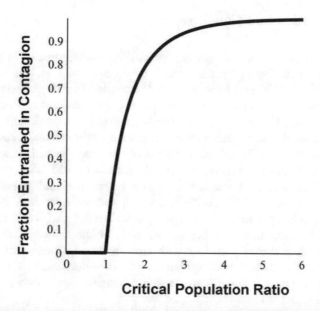

Fig. 5.4 The fraction of a susceptible population infected by a contagious process as a function of the ratio of total susceptible population to the critical population size, N/ρ, from Eq. (5.3). In 1918, some 1/3 of the total population of 1.5 billion became infected, suggesting a critical population of about 1.233 billion. By 2015, the total population has reached about 7 billion, suggesting a ratio N/ρ of 5.68. This leads to a pandemic penetrance of 0.9965, under 1918 contact probabilities

a ratio N/ρ of 5.68. This leads to a pandemic penetrance of about 0.9965, under 1918 contact probabilities.

In reality, travel patterns have much tightened in the last 100 years: jets rather than steamers. That is, in more highly connected networks, as in the nested fractals of road systems and air travel, there may be no threshold condition whatsoever. A pathogen that evolves the right balance of transmission and virulence can make it from the deepest forest or just about any backwater farm to the most globalized of population centers. The question of pandemic emergence and spread then becomes entirely dependent on local agroecological structures and processes, even as these are also interpenetrated with global circuits of capital (e.g., R.G. Wallace et al. 2014, 2015; R. Wallace et al. 2016).

The economist Douglas Almond (2006) has explored long-term consequences of the 1918 pandemic that may be relevant to our analysis. He finds

> Data from the 1960–80 decennial U.S. Census indicate that cohorts in utero during the pandemic displayed reduced educational attainment, increased rates of physical disability, lower income, lower socioeconomic status, and higher transfer payments compared to other birth cohorts.

One suspects these outcomes, in conjunction with the Great Depression, contributed to—were synergistic with—higher rates of morbidity and mortality throughout the life course of that cohort.

5.4   Discussion and Conclusions

The fatality rate of the 1918 influenza pandemic has been estimated as greater than 2.5 %, under the relatively primitive medical practices of the time (Taubenberger and Morens 2006). A 100 years later, while individual-level medical interventions are more sophisticated, for those with access, political and resource rivalries may well be greater than in 1918, at the exhausted end of a World War, particularly now under the impact of accelerating climate change and other environmental catastrophes.

The social forces driving 50 years of Cold War have now turned inward, fractioning the Soviet Union and driving increasing instability in the USA, the purported "winner" whose industrial base has crumbled under neoliberal globalization and a military economy (R. Wallace 2015, Chap. 7, and the references therein). The People's Republic of China is undergoing a wrenching shift of both economic structure and expectation in a population of 1.2 billion. Three thermonuclear-armed states are presently grappling with growing destabilization, often managed by externalizing discontent upon foreign enemies of various kinds. The attention in the USA given to minor-league "terrorist threats" arising from poorly armed Middle Eastern fanatics serves as a chilling example.

An enterprise of 7 billion people is inherently difficult to manage, in the absence of economic "farming" that regularizes both supply and demand (R. Wallace 2015). Given the accelerating pace of human-caused climate change and environmental damage, such a difficulty becomes synergistically problematic. Given the increasing inevitability of an agroecological pandemic, the fallout of acute dangers is turned into more than matters of direct mortality. Take a pathogen of a fatality rate greater than 2.5 % and—given global travel networks—universal penetrance. A central question revolves around the pathogen's expected social amplification factor. A factor of 30 is virtually a human extinction event. A factor of 10 is an unprecedented catastrophe. A factor of 2 exceeds current strategic thinking, which seems largely focused only on the pandemic event itself.

Given these uncertainties, and indeed the lack of much research at all on the social amplification of disasters, the failure of policymakers to recognize the potential for cascading loss-of-life in the aftermath of a serious, agroecologically driven pandemic calls for the most stringent application of the precautionary principle. If an action or policy has a suspected risk of causing harm to the public or to the environment, in the absence of scientific consensus that the action or policy is not harmful, the burden of proof that it is not harmful falls on those taking an action.

In the face of the potential catastrophe, it would indeed seem most prudent to begin placing draconian restraints on existing plantation and animal monocultures, the driving forces behind present pandemic emergence (R.G. Wallace et al. 2014; R. Wallace et al. 2016). To move in that direction, high-level dialog on these matters between the USA and the PRC is in order. The two countries contain or control by proxy a majority of the world's poultry and livestock monoculture. Any such effort, however, must be transverse to the present agribusiness greenwash aimed at circumventing effective regulation (Clapp and Fuchs 2009; R.G. Wallace and Kock 2012).

The political challenges, however, are daunting, as RNA viruses have powerful agribusiness interests that are in effect working on their behalf (R.G. Wallace 2009). A canonical example of political influence affecting public health, the National Rifle Association uses its influence in the US Congress to prevent the US Centers for Disease Control from examining gun violence as a public health problem, a matter involving tens of thousands of excess deaths each year (VPC 2014). Similar culturally specific patterns of command and control may be at work elsewhere, making overcoming the global corporate "virus lobby" in time enough to prevent a socially amplified agroecological pandemic a challenge of generational proportions.

Acknowledgements The authors thank H. Jack Geiger for reference to the 1962 NEJM articles, to which he was a contributor, and D.N. Wallace for useful discussions.

References

Almond, D. (2006). Is the 1918 influenza pandemic over? Long-term effects of *in utero* Influenza exposure in the post-1940 U.S. population. *Journal of Political Economy, 114*, 672–712.

Bailey, N. (1975). *The mathematical theory of infectious diseases and its applications* (2nd ed.). New York: Hafner Press.

Clapp, J., & Fuchs, D. A. (Eds.). (2009). *Corporate power in global agrifood governance*. Boston: MIT.

Drew, T. W. (2011). The emergence and evolution of swine viral diseases: To what extent have husbandry systems and global trade contributed to their distribution and diversity? *Revue Scientifique et Technique, 30*, 95e106.

Engering, A., Hogerwerf, L., & Slingenbergh, J. (2013). Pathogen - host - environment interplay and disease emergence. *Emerging Microbes and Infections, 2*, e5.

Ercsey-Ravasz, M., Toroczkai, Z., Lakner, Z., & Baranyi, J. (2012). Complexity of the international agro-food trade network and its impact on food safety. *PLoSOne, 7*(5), e37810.

Ervin, F., Glazer, J., Aronow, S., Nathan, D., Coleman, R., Avery, N., et al. (1962). I. Human and ecologic effects of an assumed thermonuclear attack on the United States. *New England Journal of Medicine, 266*, 1127–1137.

Fanon, F. (1966). *The wretched of the earth*. Boston, MA: Beacon Press.

FAO (2013). *World Livestock 2013: Changing disease landscapes*. Rome: Food and Agriculture Organization of the United Nations.

Freudenberg, N., Fahs, M., Galea, S., & Greenberg, A. (2006). The impact of New York City's 1975 fiscal crisis on the tuberculosis, HIV, and homicide syndemic. *American Journal of Public Health, 96*, 424–434.

Gandolfi, A. (2013). Percolation methods for SEIR epidemics on graphs. In V. Rao & R. Durvasula (Eds.), *Dynamic models of infectious diseases: Volume 2, Non Vector-Borne Diseases* (chap. 2, pp. 31–58). New York, Springer.

Gould, P., & Wallace, R. (1994). Spatial structures and scientific paradoxes in the AIDS pandemic. *Geografiska Annaler B, 76*, 105–116.

Jones, B. A., Grace, D., Kock, R., Alonso, S., Rushton, J., Said, M. Y., et al. (2013). Zoonosis emergence linked to agricultural intensification and environmental change. *PNAS, 110*, 8399e8404.

Kasperson, R., Renn, O., Slovic, P., Brown, H., Emel, J., Goble, R., Kasperson, J., & Ratick, S. (1988). The social amplification of risk: A conceptual framework. *Risk Analysis, 8*, 177–187.

Kentikelenis, A., King, L., McKee, M., & Stuckler, D. (2014). The International Monetary Fund and the Ebola outbreak. *The Lancet Global Health*. Published online December 22, 2014 http://dx.doi.org/10.1016/S2214-109X(14)70377-8.

Khan, S. U., Atanasova, K. R., Krueger, W. S., Ramirez, A., & Gray, G. C. (2013). Epidemiology, geographical distribution, and economic consequences of swine zoonoses: A narrative review. *Emerging Microbes Infections, 2*, e92. doi:10.1038/emi.2013.87.

Leibler, J. H., Otte, J., Roland-Holst, D., Pfeiffer, D. U., Magalhaes, R. S., Rushton, J., Graham, J. P., & Silbergeld, E. K. (2009). Industrial food animal production and global health risks: Exploring the ecosystems and economics of Avian Influenza. *EcoHealth, 6*, 58e70.

Liverani, M., Waage, J., Barnett, T., Pfeffer, D. U., Rushton, J., Rudge, J. W., et al. (2013). Understanding and managing zoonotic risk in the new livestock industries. *Environ. Health Perspect.*, 121:873e877.

Mennerat, A., Nilsen, F., Ebert, D., Skorping, A. (2010). Intensive farming: Evolutionary implications for parasites and pathogens. *Evolutionary Biology, 37*, 59e67.

Monkkonen, E. (2001). *Murder in New York City*. Berkeley, CA: University of California Press.

Parshani, R., Carmi, S., & Havlin, S. (2010). Epidemic threshold for the susceptible-infectious-susceptible model on random networks. *Physical Review Letters, 104*, 258701.

Schumer, C. (1977). Schumer reveals: Fire department death statistics may be falsified – statistical analysis demonstrates engine response time figures also doubtful. Press release dated September 26, 1977.

Taubenberger, J., & Morens, D. (2006, 1918). Influenza: The mother of all pandemics. *Emerging Infectious Diseases, 12*, 15–22.

VPC (2014). *Project Report of the Violence Policy Center*. www.vpc.org.

Wallace, D. (1994). The resurgence of tuberculosis in New York City: A mixed hierarchically and spatially diffused epidemic. *American Journal of Public Health, 84*, 1000–1002.

Wallace, D. (2001). Discriminatory public policies and the New York City tuberculosis epidemic, 1975–1993. *Microbes and Infection, 3*, 515–524.

Wallace, D., & Wallace, R. (1998). *A plague on your houses: How New York City burned down and national public health crumbled*. New York: Verso Press.

Wallace, D., & Wallace, R. (2011). Consequences of massive housing destruction: The New York City fire epidemic. *Building Research and Information, 39*, 395–411.

Wallace, R. (1990). Urban desertification, public health and public order: Planned shrinkage, violent death, substance abuse and AIDS in the Bronx. *Social Science and Medicine, 32*, 801–820.

Wallace, R. (2015). *An ecosystem approach to economic stabilization: Escaping the neoliberal wilderness*. New York: Routledge.

Wallace, R., & Fullilove, R. (2014). State policy and the political economy of criminal enterprise: Mass incarceration and persistent organized hyperviolence in the USA. *Structural Change and Economic Dynamics, 31*, 17–31.

Wallace, R., & Wallace, D. (1977). *Studies on the collapse of fire service in New York City 1972–1976: The impact of pseudoscience in public policy*. Washington, DC: University Press of America.

Wallace, R., & Wallace, D. (1990). Origins of public health collapse in New York City: The dynamics of planned shrinkage, contagious urban decay and social disintegration. *Bulletin of the New York Academy of Medicine, 66*, 391–420.

Wallace, R., Bergmann, L., Hogerwerf, L., Kock, R., & Wallace, R. G. (2016). Ebola in the hog sector: Modeling pandemic emergence in commodity livestock. In R. G. Wallace, & R. Wallace (Eds.), *Neoliberal Ebola: Modeling disease emergence from finance to forest and farm*. Berlin: Springer.

Wallace, R., Wallace, D., Ullmann, J., & Andrews, H. (1999). Deindustrialization, inner-city decay and the hierarchical diffusion of AIDS in the USA: How neoliberal and cold war policies magnified the ecological niche for emerging infections and created a national security crisis. *Environment and Planning A, 31*, 113–135.

Wallace, R. G. (2009). Breeding influenza: The political virology of offshore farming. *Antipode, 41*, 916–951.

Wallace, R. G., Bergmann, L., Kock, R., Gilbert, M., Hogerwerf, L., Wallace, R., Holmberg, M. (2015). The dawn of structural one health: A new science tracking disease emergence along circuits of capital. *Social Science and Medicine, 129*, 68–77.

Wallace, R. G., Gilbert, M., Wallace, R., Pittiglio, C., Mattili, R., Kock, R. (2014). Did Ebola emerge in West Africa by a policy-driven phase change in agroecology? *Environment and Planning A, 46*, 2533–2542.

Wallace, R. G., & Kock, R. (2012). Whose food footprint? Capitalism, agriculture and the environment. *Human Geography, 5*(1), 63–83.

Index

Printed in the United States
By Bookmasters